This edition published in 2022 by Chartwell Books, an imprint of The Quarto Group, 142 West 36th Street, 4th Floor, New York, NY 10018 USA T (212) 779-4972 F (212) 779-6058 www.Quarto.com

First published in 2018 by Motorbooks, an imprint of The Quarto Group, 401 Second Avenue North, Suite 310, Minneapolis, MN 55401 USA. Telephone: (612) 344-8100 Fax: (612) 344-8692

10 9 8 7 6 5 4 3 2 1

ISBN: 978-0-7858-4267-5

Acquiring Editor: Darwin Holmstrom
Project Manager: Jordan Wiklund
Art Director: James Kegley
Cover Illustration: Hector Cademartori
Layout: Ashley Prine

On the endpapers: (front) A 1979 Mach 1 Mustang with 429 Ram Air engine greets the morning; (rear) a 1969 Cougar Eliminator idles at sunset. *Jerry Heasley*
On the frontis: Technical editor Jerry Titus was the driver for *Sports Car Graphic*'s test of a 1964 Cobra. Titus would soon join Shelby as a team driver. *Archives / TEN: The Enthusiast Network Magazine, LLC*

Printed in China

THE COMPLETE BOOK OF CLASSIC FORD AND MERCURY MUSCLE CARS 1961–1973

DONALD FARR

NAS 042
WASH 63

CONTENTS

Acknowledgments

While my nearly four-decade automotive journalism career has included stints at *Super Ford* and *Musclecar Review* magazines, I'll admit that I've mainly focused on Mustangs. No shame there, only an acknowledgement that I required much help outside the Mustang realm while writing this book, which has given me new appreciation for the car owners and enthusiasts who specialize in knowing every little detail about their favorite Ford or Mercury. Many of them responded to my pleas for information, specifications, and road-test articles, and I can't thank them enough for taking the time to respond to emails or take phone calls to supply info. So, with hope that I am remembering everyone, the best I can do is acknowledge them here: Marty Burke, Austin Craig, Mike Eaton, Bill Hamilton, George Huisman, Phil Jamieson, Rick Kirk, John Kranig, Steve and Tayna Mank, Ed Meyer, Tim Orick, Dan Parson, Bob Perkins, Barry Rabotnik, Jim Smart, and Jim Wicks. Also, Bill Barr took the time to talk with me about his work at Ford Engine Engineering, where he was involved in the development of the 428 Cobra Jet.

These days, there are clubs and registry websites for nearly every popular Ford and Mercury muscle car. While researching this book, I clicked through many of them, including the Mustang 428 Cobra Jet Registry, 429 Mustang Registry, Cougar GT-E Registry, Cougar Eliminator Registry, Boss 302 Registry, and the Shelby American Automobile Club. Some members went beyond the call of duty by responding to my numerous requests for information. They included Bob Mannel (Fairlane Club of America), Mark Reynolds (Galaxie Club of America), Kirk Dillery (Mercury Marauder Club), and Rob Day (Cyclone/Montego/Torino Registry), along with David Wagner, Chuck Beason, Dick Harrington, and Mac McCray from the Falcon Club of America. And on the rare occasion when I couldn't find what I needed at the informative Cougar Club of America website, Jim Pinkerton, Don Skinner, Bill Quay, Dave Wyrwas, Phil Parcells, and Mitch Lewis came through with information from their areas of specific model expertise.

Another invaluable website was the Old Car Manual Project at www.oldcarbrochures.com, which features sales brochures from American auto manufacturers. The Ford and Mercury brochures from the 1960s and 1970s provided many of the engine specifications, power ratings, and high-performance, model-specific, equipment listings for this book.

For perspective, each chapter lead includes an anecdote from someone who experienced Ford and Mercury muscle cars either new or almost new. Many thanks to "Animal Jim" Fuerer, Richard Adis, Ricky Ward, Herb Gordon, Richard West, and Hunt Palmer-Ball for sharing their stories.

Ford and Mercury muscle cars were produced in relatively small numbers, which certainly adds to today's interest, appeal, and collectability. Thanks to Kevin Marti at Marti Autoworks (www.martiauto.com), we have access to Ford's production data for 1967 and later vehicles. Much of Kevin's numbers-crunching can be found at the aforementioned websites, within "Marti Reports" for individual cars, or in Kevin's Mustang and Cougar *By the Numbers* books, but he also dug a little deeper for me to pull out specific info for this book. I'm fortunate to count Kevin among the many Ford friends I've made along the way.

Another longtime pal and former co-worker from decades ago is Mike Mueller, who graciously shipped books, copied magazine articles, supplied photos, and provided contact info that contributed to this book. I'm sure the digging through files and time at the copy machine took time away from his own automotive book projects, and for that I'm deeply grateful.

When it came to photos, the ever-popular Mustangs were not a problem. Try finding photos of factory original Cougar and Cyclone muscle cars, however, produced in small numbers to begin with and most succumbing to the ravages of time and abuse. Over my career, I've crossed paths with many automotive photographers. You would not be holding this book in your hands if it weren't for photographic contributions from Dale Amy, Juan Lopez-Bonilla, Eric English, Cam Hutchins, David Newhardt, Al Rogers, and Jim Smart, along with Alex Yankovich, who provided photos from the Mecum Auction files. Just as I was starting on this project, longtime friend and ace muscle car photojournalist Tom Shaw passed away. My thanks to his family, in particular son Austin Shaw and daughter Robin Mason, for allowing me to use Tom's photography in this book. And when my photo quest for some of the rarer Ford and Mercury muscle cars came up short, I put in a call to my longtime cohort Jerry Heasley, whose, "Oh, I've got that one" responses were joy to my ears.

I was also fortunate for the opportunity to once again work with Thomas Voeringer, the archivist at The Enthusiast Network, who maintains the massive photo archives from the former Petersen Publishing magazine empire, which included *Motor Trend*, *Hot Rod*, *Car Craft*, and *Sports Car Graphic*. Thomas dug up contact sheets from the old magazine road tests so I could choose vintage photos for this book, which I felt was important to show the cars as they were delivered new.

GALAXIE 500
MICH 65
TP-9946
WATER-WINTER
WONDERLAND

INTRODUCTION

"There's something about setting back into the deep bucket seats in that all-black interior basking in the soft green light from the gauges and listening to the engine noises as you glide and bump along in a tunnel of mercury vapor street lights. The chrome Hurst shift lever picks up the light from the radio dial, the tach and speedo boom out loud and clear right at you, and you feel good."

Motor Trend's A. B. Shuman captured the essence of the muscle car experience in his test-drive report about the 1970 Boss 302 Mustang. From Woodward Avenue to Van Nuys Boulevard to the neon-lit drive-ins in small towns all across America, young men cruised the city streets and two-lane blacktops in their 427 Galaxies, Cobra Jet Cougars, and Mach 1 Mustangs. Back then, these high-horsepower factory supercars were daily-drivers for school or the job at the local textile mill, but they were also used for Saturday night fun, either to attract girls or to challenge the 440 Plymouth at the red light. With cheap high-octane fuel available at many street corners and an emerging performance parts industry feeding the demand for even more power, muscle cars were a way of life for many hot-rodding thrill seekers in the 1960s and early 1970s.

Ford Motor Company tapped into the market with its Total Performance marketing campaign in the early 1960s, promoting racing and high-performance cars like never before. Big Galaxies and Mercury Marauders eventually gave way to smaller, lighter Fairlanes and Cyclones before the pony car craze, spawned by the Mustang, took hold to produce Super Cobra Jet Mach 1s and Boss 429s. *The Complete Book of Classic Ford & Mercury Muscle Cars: 1961–1973* documents the Blue Oval's contributions to the exciting muscle car era.

It was a special time in American automotive history—and an exciting time for those of us who lived through it.

For 1965, the 427 was available in all Galaxies, including the 500 convertible. *Dale Amy*

Prepped for Performance

Throughout the muscle car era, the more powerful engines under the hood demanded heavier-duty equipment in other areas to withstand the hard-use demands expected from high-performance cars. Here are some of the components that were added to Ford and Mercury muscle cars from 1961 to 1973.

Heavy-duty suspension: Ford and Mercury used several names to describe their upgraded suspensions—Special Handling Package, Competition Suspension, Heavy-Duty Suspension, etc. In most cases, the packages included higher-rate springs, stiffer shocks, and larger front sway bar, with specifications varying depending on the engine. Larger 15-inch wheels were sometimes part of the upgrade and a rear sway bar was added to some suspensions starting in 1969.

Toploader four-speed: Ford introduced the Toploader in 1964 as a replacement for the Borg Warner T-10. Internal parts were installed through the top of the case (thus the "Toploader" description), which provided a stronger case than the earlier side-loading transmissions. Toploaders were fully synchronized except for reverse, allowing for quicker shifts and the ability to downshift with the car moving. Built in several lengths (depending on the vehicle), Toploaders came with 28- and 31-spline output shafts, with the larger diameter found in 427, 428, and 429 models.

Hurst shifter: In 1970, Ford began installing a Hurst shifter with T-handle in four-speed performance cars. It was the shifter arm only; the linkage was production Ford.

Nine-inch rear end: Big power combined with full-throttle shifting placed a lot of stress on the drivetrain, especially the rear end, and Ford had just the answer with its 9-inch rear axle, which debuted in 1957 and is today still recognized as one of the strongest rear ends of all time. It was used on nearly all high-performance Fords from 1961 to 1973 in several widths (depending on model) and with additional heavy-duty components added as torque and horsepower increased, including 31-spline axles, nodular iron differential case, and larger axle bearings. A limited-slip differential was optional, including the later Traction-Lok and Detroit Locker. Typical gear ratios for muscle cars ranged from standard 3.25 or 3.50 to optional 3.91, 4.11, or 4.30.

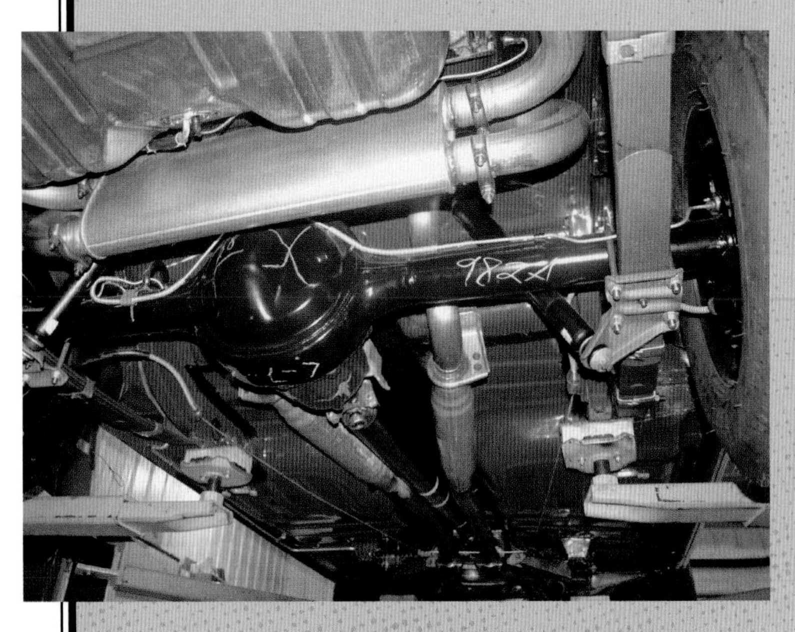

top: When powered by the stout R-Code 428 Cobra Jet, the 1970 Mach 1 came with additional heavy-duty components, including Competition Suspension and nine-inch rear axle, plus Hurst shifter and staggered rear shocks when equipped with four-speed.

opposite left: Ford's extra-strong nine-inch rear end was installed in most Ford and Mercury muscle cars. This one is mounted with staggered rear shocks—one in front and one behind the rear axle—as used on 1968–73 four-speed cars to reduce wheel hop during hard acceleration. *Donald Farr*

opposite right: Several 1969–71 Ford and Mercury muscle cars with four-speed used an electronic rev limiter to prevent over-revving. *Donald Farr*

Staggered rear shocks: Beginning in 1968 with the 428 Cobra Jet, Ford began employing a staggered rear shock arrangement on cars equipped with four-speed transmission. By placing one shock in front of the rear axle and one behind, axle wrap-up and the resulting wheel hop was reduced under hard acceleration. Exceptions included the 1969 Talladega, which had staggered shocks and automatic transmission, and the 1971–73 Mustangs with Competition Suspension, which were known to have staggered shocks with automatic transmission,.

Rev limiter: In an attempt to reduce warranty claims, Ford added an electronic governor to a number of 1969–71 high-performance models with four-speed transmission. Designed to prevent over-revving, the rev limiter was wired into the ignition system to create a misfire at a pre-set rpm, either 5,800 (428 Cobra Jet) or 6,150 (Boss 302/429 and 429 Cobra Jet). Mounted on the inner fender, the rev limiter was wired into the distributor to coil wiring and was easily disconnected. Most were removed and thrown away.

Four-bolt mains: Ford V-8 engines secured their five main bearing caps with two bolts, which was sufficient for normal driving conditions. However, muscle cars weren't driven normally, typically experiencing high rpms and speed-shifting, so their main bearings were often strengthened with two extra bolts. Most 427s utilized cross-bolted mains, while later engines, notably the Boss 302/351 and some four-barrel 351 Clevelands, received extra bolts on the middle three or all five main bearing caps.

TENN
61
19·7649

CHAPTER ONE

THE BIG BODIES

1961-1970

In February 1963, Jim Feurer was 22 years old and bringing home good money from his job in Chicago with Western Electric. An admitted "gear head" with a penchant for fast cars, Feurer placed an order for a 409 Chevy but canceled it when delivery was delayed. A subsequent deal for a Max Wedge Dodge fell through when the dealer wouldn't accept Feurer's trade-in. Then a friend who worked at the local Ford dealership called. "He had just heard about the 427 for the Monterey," Feurer recalls. "I ordered it that day, a black two-door sedan with Super Marauder 427 and 4.11 gears." Feurer took delivery on March 23 and immediately started street racing. He lost only once with his 427 Merc—to a 427 Galaxie, he said. When a friend noted that the big Monterey launched "like an animal," Feurer gained a nickname. Throughout his future drag racing career, he was known as "Animal Jim."

At the beginning of the muscle car decade, the 1961 Starliner with the 401-horsepower 6V Thunderbird Special was the hot setup. *Jerry Heasley*

Feurer was typical of the early 1960s speed addict. Single, carefree, and flush with disposable income, young men like Feurer craved the big-engine, big-body cars from Detroit. Technically, the term "muscle car" wasn't coined until Pontiac dropped a 389 into the midsize 1964 Tempest and called it the GTO. But if the description had been used earlier, it surely would have described the brawny 1961–64 full-bodied Fords and Mercurys with high-performance 390, 406, and 427 engines. Although big in size and heavy in weight, the 400-horsepower ratings satisfied the cravings for tire-squealing acceleration and adrenaline-inducing top speeds.

The 360-horsepower 352 High Performance engine for 1960 lit the fuse. For the next four years, big Fords with solid-lifter 390s, six-barrel 406s, and dual-quad 427s exploded onto the American performance car scene with ever-increasing displacement and horsepower, all the way to 425 by 1963. Ford was also locking horns with General Motors and Chrysler in NASCAR and professional drag racing, which resulted in a "Win on Sunday, sell on Monday" mentality and the company no longer participating in the Automobile Manufacturers Association's (AMA) racing ban. To keep things from getting out of hand, sanctioning bodies established a homologation rule that mandated a certain number of engines and cars must be produced for sale through dealerships. In 1962, Ford's marketing slogan transitioned from "Extra Lively" to an all-out "Total Performance" assault.

It was the perfect storm for young men like Jim Feurer.

above: Jim Feurer was proud as a new poppa when he came home with a brand-new 427-powered 1963 Monterey sedan with "breezeway" rear window. *Jim Feurer*

opposite top: Ford's high-performance 390s were the ideal powerplants for the 1961 Galaxie Starliner, a two-door with a sporty fastback roofline. *Jerry Heasley*

opposite left: In 1960, Ford's 352 High Performance was rated at 360 horsepower thanks to its solid-lifter drivetrain, header-style exhaust manifolds, and a Holley four-barrel on an aluminum intake. It would mark the beginning of a three-year displacement and horsepower spurt for the FE engine. *Donald Farr*

opposite right: When equipped with six-barrel Holley carburetion, either from the factory or installed by the dealer, the horsepower rating for the solid-lifter 390 jumped to 401. *Jerry Heasley*

1961 Galaxie Solid-Lifter 390

Hot Rod called it "Ford's Hot Stocker." In the December 1960 issue, editor Ray Brock glossed over Ford's styling updates for the 1961 Fords and headed straight for the meat—the 375-horsepower 390 Thunderbird Super V-8. "More displacement and increased horsepower promises to make Ford enthusiasts happy," he predicted.

In 1960, Ford wet the whistle for race and driving enthusiasts with the solid-lifter 352, rated at 360 horsepower. For 1961, Ford expanded the 352's bore and stroke to create the 390 Thunderbird Special, a primarily passenger-car engine with 300 horsepower. By November 1960, Ford had unleashed the Thunderbird Super 390 with 375 horsepower as the replacement for 1960's solid-lifter 352. As Brock pointed out in *Hot Rod*, the high-performance 390 was much different than the mundane passenger-car version, using a stronger block, specially selected pistons, smaller combustion chambers for a 10.6:1 compression ratio, high-lift cam with solid lifters, dual-point distributor, free-flowing cast-iron headers, and an aluminum intake manifold with a Holley four-barrel. While visiting Dearborn, Michigan, to preview the 1961 Fords, Brock reported that a hardtop equipped with the 375-horse 390 had topped 159 miles per hour on Ford's test track.

Impressive—until Ford added another late addition to the engine lineup. The 6V Thunderbird Special had essentially the same solid-lifter powerplant as the Thunderbird Super 390 but with an aluminum intake that mounted a trio of Holley

Starliner

two-barrel carburetors. Rated at 401 horsepower and available in early 1961, the 6V 390 was Ford's first factory engine rated at more than 400 horsepower. The timing was critical; Ford wanted to make sure the 401 horses were legal for January's NHRA Winternationals. Many of the units were supplied in the trunk for dealer installation, but a few full-size Fords rolled off the assembly line with the six-barrel induction.

The Thunderbird Super 390 and 6V Thunderbird Special were offered only with the three-speed manual transmission, with or without overdrive, although a Borg-Warner four-speed became available late in the model year. Both 390s were complete performance packages; cars so equipped got heavy-duty springs and shocks, a larger 3/8-inch fuel line, wider station wagon front drum brakes, 15-inch wheels, a larger diameter driveshaft, and a four-pinion differential with 3.89:1 gear ratio for the three-speeds or 4.10:1 for overdrive. Power brakes and steering were not available.

1962 Galaxie 406 6V

Ford ended 1961 on a performance high note with 375- and 401-horsepower versions of the solid-lifter 390 for full-size Fords. The two engines continued into the 1962 model year but with new names—390 High Performance and 390 Super High Performance. However, both would be discontinued around January 1962. And for good reason—Ford had a couple of tricks up its sleeve for midyear 1962 introduction.

The FE: Displacement and Power

With passenger cars growing larger and heavier in the late 1950s, combined with the public's growing thirst for racing and performance, Ford needed a larger displacement companion for the Y-block, introduced in 1954 with 239 cubic inches and maxed out at 312 cubic inches by 1956. Coinciding with the introduction of the 1958 Edsel, Ford introduced a new engine, also based on the Y-block's strong design with the skirt extending below the crankshaft centerline but with the capacity for a larger bore. Ford identified the powerplant as the "Interceptor" and, the following year, as the "Thunderbird Special," but it became better known by its internal code name—FE, for "Ford Edsel."

The FE debuted in 1958 with a 4.00-inch bore and 3.30-inch stroke to create a 332-cubic-inch engine with either two- or four-barrel induction. A 352-cubic-inch Interceptor Special with 3.50-inch bore was also available for the Fairlane 500 and station wagons. The FE was exclusive in the new Edsel as either 361 or 410 cubic inches. The FE's more than 400-cubic-inch potential would prove beneficial in the coming years.

As a newer design, the FE also incorporated a number of improvements over the earlier Y-block, including easier serviceability, wider bore spacing, a lighter valvetrain, and larger intake ports. In the late 1960s, the FE big-block would serve as Ford's workman-like engine, not only for trucks and four-door family sedans but also for performance duty as displacement increased from 352 to 390, 406, 427, and 428. From six-barrel 406s and dual-quad 427s to the later 428 Cobra Jet, from numerous NASCAR victories to victory lane at LeMans, the FE engine would serve as the big-inch foundation for Ford's Total Performance commitment.

Following Chevrolet's introduction of a Super Sport package to showcase its 1961 Impala with the 409, Ford fired back in February 1962 with the Galaxie 500/XL with bucket seats, console, and special interior trim. The sales brochure explained that XL stood for "Extra Lively," but it was also the perfect platform for Ford's other midyear surprise—the 406.

The result of a .08-inch bore increase for the 390, the 406 was offered as a either a 385-horsepower High Performance with a single Holley 600 cfm four-barrel or a 405-horse Super High Performance with three Holley two-barrels, with both engines using solid-lifters, aluminum intake manifolds, and open-element air cleaners. For normal driving, the 6V engine operated on the 330 cfm center carb; at wide-open throttle, the progressive linkage opened the pair of outboard two-barrels rated at 350 cfm each. The full-throated sound of wide-open six-barrel induction was music to a young hot-rodder's ears.

above: An oval air cleaner hid the trio of Holley two-barrels that boosted the 406's horsepower to 405. *Eric English*

top: Except for the station wagon, the 401-horsepower 406 6V was available in all 1962 big-body Fords, including the convertible, made all the more sinister with dog-dish hubcaps. *Eric English*

Like the previous high-performance 390s, the 406 block was stronger; a few even received cross-bolted mains and higher-strength main bearing webs. The 406 also made use of new flat-top pistons, heavy-duty connecting rods, larger 2.09/1.66-inch intake/exhaust valves, and high-flow exhaust manifolds. Ordering one of the 406s automatically equipped a big Ford with the heavy-duty suspension, fade-resistant drum brakes, higher capacity radiator, and 15-inch wheels. Also mandatory was a manual-shift transmission, either three-speed (with standard column shift or optional floor shift), three-speed with overdrive, or the new Borg-Warner T-10 four-speed. Rear axle ratios up to 4:11.1 were optional to replace the standard 3.50s.

Total Performance

In 1962, Henry Ford II made it clear: "The AMA ban on racing is null and void within Ford Motor Company." Racing and performance activities, which had been unofficially handled through back-door channels, were now out in the open. Ford was free to unleash its engineering and marketing forces on a public that was craving speed for both street and track.

Since mid-1957, Ford had loosely abided by an Automobile Manufacturers Association resolution that prevented American auto companies from participating in racing, including the publicizing of race results and advertising speed-related cars. Chrysler and General Motors had been even less compliant as they continued to quietly develop performance equipment, putting Ford behind the curve. Henry Ford II had recently been re-elected as president of the AMA when he advised the organization, "The resolution adopted in the past no longer has either purpose or effect. Accordingly, we are withdrawing from it."

Within months, Ford was actively transitioning its advertising slogan from "Lively One" to "Total Performance." In an April 1963 statement, Ford's Lee Iacocca made the first reference to the new term: "Performance has been integral in the long history of Ford Motor Company. We at Ford believe in performance, because the search for performance—Total Performance—made the automobile the wonderfully efficient, pleasurable machine it is today—and will make it better tomorrow."

As demonstrated in everything from magazine ads featuring tire-smoking Galaxies to sales literature promoting four-door sedans, Ford's Total Performance commitment became top priority. In the coming months and years, Ford's presence would be felt in racing, everywhere from the NASCAR ovals and drag racing quarter miles to the victory lane at LeMans. For homologation purposes, professional racing required production street cars, which in turn benefitted from the "Win on Sunday, Sell on Monday" performance marketing, resulting in Galaxies and Marauders powered by race-inspired 427s and, later in the decade, Mustangs such as the Boss 302 and Boss 429.

MAYFIELD FIRE STATION Nº 5
Ford owners: discover how much your old flame has changed! Slide inside and feel a new brand of Total Performance—refined and perfected in events at Atlanta, Daytona, Riverside! Quit answering false alarms... discover a real change...in a Super Torque Ford.
TRY TOTAL PERFORMANCE FOR A CHANGE!
FORD
Falcon · Fairlane · Ford · Thunderbird

Ford's Total Performance marketing campaign was in full blossom by the time the 1963 ½ Galaxie fastback arrived with available 427 power. By pulling out of the Automobile Manufacturers Association racing ban, Ford was free to promote racing and performance. *Donald Farr*

Available in all models except the station wagon, the new 406s were particularly well suited to the 500/XL's sporting demeanor, although some found their way into lighter two-door sedans, which were affectionately known as "box tops" for their squared-off roof.

Naturally, the car magazines zoomed in on America's latest performance engines, including *Car Life,* with a March 1962 comparison of the Ford 406, Chevy 409, and Chrysler 413. Installed in a Galaxie 500, the 405-horse Ford impressed the editors: "Our test Galaxie 406 would romp up to 6,000 rpm in each of the four gears so fast that it literally made our head swim. There is so much punch, even in third and fourth gears, that it really doesn't feel much different from WOT in first and second."

Motor Trend described the sound: "The triple two-throat carburetion and low-restriction mufflers give a throaty roar like a Gold Cup hydroplane when you punch the throttle. Wonderful! The mechanical progressive throttle linkage (where you run on only the center carb until about 2/3-throttle opening) seems very efficient. You get the usual slight hesitation (and maybe half a cough) when all the barrels are plopped open at once at low rpm. But overall the throttle response is good."

1962–63 Super Marauder 406

With the horsepower wars fully underway, Mercury joined the fun in 1962 with an S-55 package for the entry-level Monterey hardtop and convertible. Part of Mercury's "Special" series that also included the Comet S-22 and Meteor S-33, the full-size S-55 mirrored Ford's Galaxie 500/XL by packaging bucket seats, console with floor shift, distinctive door panels, and chrome interior trim, all for $520 over the standard Monterey Custom.

Ford called it the 406 Super High Performance. Mercury's name for the 405-horse 406 6V was Super Marauder 406.

By late in the 1962 model year, and like its full-size Ford cousin, the S-55 could back up its sporty credentials with a pair of high-performance 406-cubic-inch engines. Mercury called them "Marauder V-8s"—a 385-horsepower Marauder 406 with a four-barrel or a 405-horse Super Marauder 406 with a trio of inline two-barrels. In May 1962, Ford's product engineering office signed off on the two 406 engines for the Monterey, limiting them to four-speed only and no power steering, power brakes, or air conditioning. Buyers looking for sporty roadwork combined with Mercury luxury paid an extra $321 for the four-barrel 406 or $406 for the three two-barrels. Due to late availability and additional cost, only 125 or so full-size Mercurys were sold with 406 engines for 1962.

After *Motor Trend* put a Super Marauder–powered 1962 S-55 convertible to the test, it said, "When the road clears and the three two-barrel carburetors are punched open, you've got a tiger by the tail." *Archives / TEN: The Enthusiast Network Magazine, LLC*

For an October 1962 test, *Motor Trend* got its racing-gloved hands on a 1962 Monterey S-55 convertible powered by the six-barrel 406 with four-speed and 3.50:1 gears. Weighing more than 4,200 pounds, the convertible wasn't the ideal body style for performance, but the *Motor Trend* editors were still impressed by the top-of-the-line 406: "Does it go? Answer: Yes, in every respect."

Noting that quarter-mile times were further bogged down by the weight of two testers on board, the big Mercury clocked in at 16.5 seconds at 94 miles per hour. "It was in honest street trim," noted the text, "and just a little tinkering would have raised the speed past the century mark and lower elapsed times (ETs) substantially."

The *Motor Trend* reviewers were impressed by the engine's tractability, stating,

> "The old theory about a factory hot rod being sensitive and hard to manage in traffic can be tossed right out the S-55's window. We accelerated from as low as 12 mph in fourth gear. Cracking all three carbs open at that speed caused some stumble but no permanent stalling. It has so much torque that it is virtually a two-gear automobile."

Car Life tested the same convertible and managed to improve quarter-mile performance, throttling the big Merc to a 16.1-second clocking at 85 miles per hour. Overall, they were impressed: "While the engine and components are available in other Mercurys, the S-55 makes for a sort of double-dipped luxury; you have one form of luxury in the gobs of power and speed so readily at hand, and another form in the richness of the appointments. It's sort of like having layer cake for dinner and chocolate ice cream for desert."

The optional 406 powerplants continued into 1963, but only ninety-three were sold, mainly because, at midyear, the 406 was replaced by a newer, even larger displacement version of the FE engine.

1963½–64 Galaxie 427

With Ford's commitment to Total Performance, the gloves came off and Ford came out swinging. Halfway through the 1963 model year, the Ford man on the street—and on the racetracks—started reaping the rewards.

The 1962 engine lineup rolled into 1963 with the previous year's 385-horsepower 4V/406 High Performance and 405-horse 6V/406 High Performance. But with Ford's withdrawal from the AMA ban, the doors were thrown open for two performance enhancements at midyear 1963. They were deemed so important that Ford issued a revised sales brochure—dated December 1962, just four months after the introduction of the 1963 models—to add a new fastback body style and a pair of even larger displacement engines.

"The XL Sports Hardtop has the fresh new slipstream look," said the updated copy. In truth, the aerodynamic fastback roofline was designed to slip through the air on the NASCAR high-speed ovals, although the sleek profile also fit the sporty image that appealed to younger, mostly male buyers.

Available as a Galaxie 500 or 500/XL (for "extra lovely," said the revised copy), the new Sports Hardtop provided the supporting image for Ford's other pleasant midyear surprise: a 427-cubic-inch replacement for the 406. Equipped with 406 cylinder heads (later to be identified as "Low Risers"), the 427 was offered as a Q-code four-barrel

The Galaxie Sports Hardtop and the 427 big-blocks arrived almost simultaneously at midyear 1963, providing the big-body with a one-two punch for Ford's Total Performance marketing. *Mike Mueller*

The Magic Number: 427

Even as a two-door hardtop, the 1964 Marauder looked more like a big luxury car. But with the 425-horsepower 427 under the hood, it could blow away 409 Chevys and Max Wedge Dodges. *Matt Magnino/Mecum Auctions*

For American auto manufacturers in the early 1960s, NASCAR was the means for promoting technical innovation. Not only did racing generate publicity, but it also provided high-profile testing grounds for performance and durability. In the quest for faster speeds on the ovals, the manufacturers increased engine displacement—Ford from 352 to 390 and 406 cubic inches for its FE big-block, Chevy from 348 to 409, and Chrysler from 331 to 392 and 413. When NASCAR established a 428-cubic-inch cap for the 1963 season, the manufacturers finally had a target.

By increasing the 406's bore another .10 inch, up to 4.23 inches, Ford arrived at 425 cubic inches, a safe 3 cubic inches below NASCAR's limit. For marketing purposes—and to top Pontiac's 421 and Chrysler's 426 while matching Chevy's rumored new big-block—the displacement was "rounded up" to 427, a number that would become legendary in Ford racing and performance history.

For the street, the 427 first appeared as an option for 1963½ Fords and Mercurys. Two versions were offered: one with a single four-barrel for 410 horsepower and the second with dual four-barrels for 425 horsepower. They were essentially race engines with cross-bolted mains, stronger connecting rods, solid-lifter camshafts, and aluminum intakes with Holley carbs.

The 427 was offered as an option from 1963½ to 1967, primarily for full-size Fords and Mercurys but also in the 1966–67 Fairlane and 1967 Comet—and of course in Carroll Shelby's 427 Cobra. The 427 evolved over time, mainly for competition purposes. An oiling weakness was addressed at midyear 1965, resulting in a new "side-oiler" block that reversed the priority of oiling, first supplying oil to the main bearings instead of the valvetrain, like earlier "top-oiler" blocks.

During the 427's production lifetime, the height of the intake manifold above the cylinder head distinguished the three different versions, not counting the race-only Tunnel Port and single overhead cam (SOHC). The taller the intake, the more direct the path from the carburetor to the intake valve, thus the ability to produce more power.

Low Riser: The first 427 cylinder head was a holdover from the earlier 406 with 2.34x1.34-inch ports, 2.04-inch intake valves, and 1.66-inch exhaust valves. In March 1963, the intake valves were enlarged to 2.09 inches. Low Riser 427s were used in the 1963½ to early 1965 Galaxies and Mercurys.

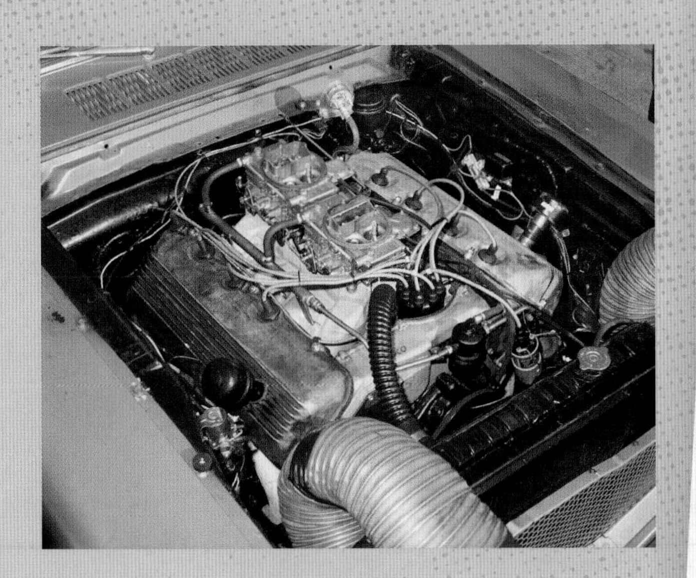

above: In its highest competition level, the 427 was developed with single overhead cam (SOHC) heads. It was quickly outlawed by NASCAR and used primarily for drag racing. The "Cammer" was never installed in a production Ford vehicle. *Archives / TEN: The Enthusiast Network Magazine, LLC*

right: Ford's 427 advertising aptly promoted the dual-quad Holleys: "With eight pipes in the organ, what noble music it makes!" *Donald Farr*

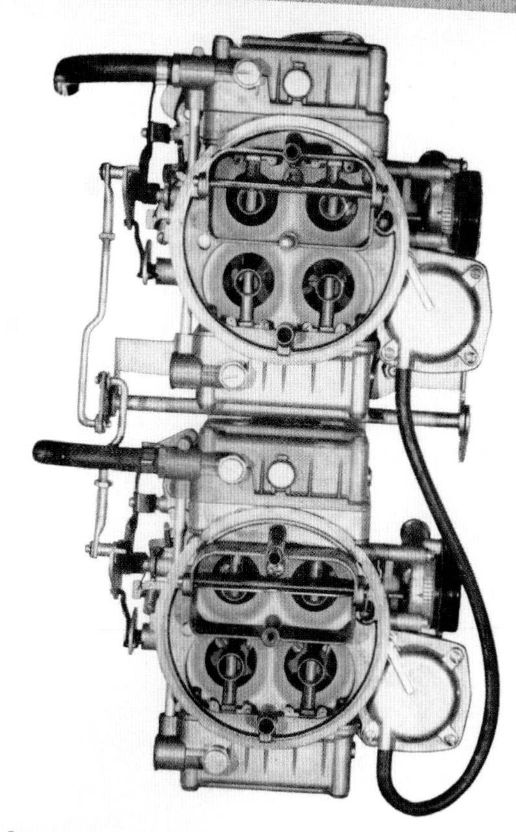

High Riser: Developed in late 1963 for NASCAR and drag racing only, these heads featured tall 2.72x1.34-inch intake ports and a matching intake manifold that positioned the carburetor well above the ports. Because the tall intake would not clear production hoods, reverse "teardrop" bubble hoods were created for the competition-only lightweight Galaxies and Fairlane Thunderbolts.

Medium Riser: When NASCAR outlawed the teardrop hood, Ford engineers went back to their drawing boards to create a new head for 1965 with smaller intake ports, 2.06x1.38 inches, but with an improved internal design, allowing them to flow nearly as well as the High Riser. The Medium Riser 427 found its way into late 1965–67 Galaxies and 1966–67 Fairlanes and 1967 Comets.

In its original solid-lifter form, the 427 made a final appearance in the 1967 Galaxie and Fairlane/Comet. For 1968, the 427 was tamed with a single four-barrel and hydraulic cam, making 390 horsepower for the Cougar GT-E. Early 1968 sales material also lists the W code 427 for the Mustang and Fairlane, but none were built.

The "flying Thunderbird with checkered flags" emblem on the front fenders gave fair warning to anyone thinking about wrangling with a 427 Galaxie. *Donald Farr*

with 410-horsepower or an R-code with two four-barrels for 425 horsepower, matching Chevrolet's rating for its dual-quad 409.

Other than their aluminum intakes and Holley carburetion, the two 427 engines were identical, with a solid-lifter camshaft, heavy-duty rod bolts, a dual-point distributor, and high-flow, cast-iron exhaust manifolds that dumped into 2-inch pipes and dual mufflers. Transmission choice was limited to the Borg-Warner T-10 four-speed, followed by the stout 9-inch rear end with 3.00, 3.50, or 4.11 gearing in a four-pinion, non-locking differential.

Like the earlier 390 and 406 models, Galaxies with the 427 were packaged with heavy-duty equipment, including springs and shocks, wider front drum brakes, and 15x5½-inch Kelsey-Hayes steel wheels. Full wheel covers with spinners were supplied on the XL models; others received plain "dog-dish" hubcaps on painted wheels. Gold "427" badges warned the competition.

While the 500/XL supplied sporty good looks, smart street racers ordered the 427 in the inexpensive, lightweight sedan. Having rubber floor mats and even lacking AM radio and sun visors, the stripped-down two-door didn't weigh much more than the lightweight 427 Galaxies supplied to racers.

The 427 Galaxie Registry estimates that Ford produced 4,895 427 Galaxies for 1963 and another 3,104 for 1964, bringing total production to 7,999.

By the time the car magazines were able to test the new 427 Galaxie, the 1964 models were out with their mildly revised styling. Otherwise, mechanicals for 1964 remained the same for both the single four-barrel and dual-quad 427s, along with the mandatory heavy-duty hardware. At some point, after reports of axle failures, the 28-spline axles were replaced by 31-spline versions.

Motor Trend tested a 1964 Galaxie 500 two-door hardtop with the 425-horse 427, describing it as "a superior road performer on one carb and a brutal tiger when we slammed in the extra carb for fast passing." With 4.11:1 gearing, the two-ton

1963½–65 Super Marauder 427

In February 1963, Lincoln-Mercury issued a press release:

> "Two new hardtop models with fastback rooflines and two high-performance engine options will be available as midyear additions. The two-door fastbacks—Marauder and Marauder S-55—are being offered in Mercury dealerships in early March. The optional 427-cubic-inch V-8 engines, which replace the 406 V-8s, will develop up to 425 horsepower."

Like the big Fords, Mercury entered 1963 with the carryover Marauder and Super Marauder 406 engines. February's announcement about 427 availability put Mercury on equal performance footing with Ford, only with Mercury's upscale twist. Contrasting Mercury's smooth-ride image, the 427-powered cars received a heavy-duty suspension with stiffer springs and shocks along with larger anti-roll bar.

Car Life announced the 427 Marauder by describing the dual-quad big-block as an engine with "positively brutal horsepower." With its test car, a 4,155-pound S-55 two-door with a four-speed and 4.11 gearing, the editors compared the 427's 15.1-second quarter-mile performance to a previously tested 406 S-55 with a four-speed and 3.56 gears, which ran best at 16.1. "The 427 doesn't take no for an answer," the magazine said, "and the Super Marauder S-55 is startling quick. Having all that power on tap under one's right foot makes passing a pleasure and hill-climbing fun."

The 427-powered Mercury was a rarity for 1963. Only eighty-three were equipped with the 427 for 1963½—twenty-five with the 410-horsepower four-barrel Marauder and fifty-eight with the 425-horse Super Marauder V-8.

The 427 continued as a Mercury full-size option in 1964, but only sixty-four were sold, twenty-two with the single-four engine and forty-two with the dual-quad R code. For 1965, the number dropped even further with Mercury offering only the 425-horse engine. Just seven were produced.

With that, the 427 era ended for the big Mercurys. For 1966, the race-inspired big-block was gone, replaced by a smoother 345-horsepower 428 as Mercury shifted toward luxury Park Lanes, not all-out performance cars such as the 427 S-55.

The price
is medium...
the action
maximum...
the car is
Mercury
Mercury
Marauder
Pikes Peak Champion
The name is the tip-off . . . Marauder! This is an action car. Looks it. Acts it. A 390 cu. in. V-8 is standard. Optional
engines range up to an 8-barrel, 427 cu. in. V-8—the newest edition of the engine that set a new world's stock-car
record in the most recent Pikes Peak Climb. Choose from six Marauder models—2-door or 4-door hardtops.
Or, if you prefer, the same performance is available in Mercurys with Breezeway Design (the rear window
opens for ventilation). See both at your Mercury dealer's. LINCOLN-MERCURY DIVISION Ford MOTOR COMPANY
'64 Mercury
No finer car in the medium-price field

Mercury promoted the 1964 Marauder by praising the available 427 as the engine that set a new stock car record in the Pike's Peak International Hill Climb. *Donald Farr Collection*

Ford covered the quarter mile in 15.4 seconds at 95 miles per hour. "Although it can be docile," *MT* continued, "the 427 comes on with a roar and smack-in-the-back acceleration that can only be termed fierce at any speed. It isn't for the faint of heart—it's a man's car through and through—and it takes a man to get the most out of it."

above: The big and heavy 427-powered Galaxie wasn't a corner-carver, but that didn't stop *Motor Trend* from checking out the handling in the Southern California mountains. *Archives / TEN: The Enthusiast Network Magazine, LLC*

opposite: Ford promoted the 427-powered 1965 Galaxie as a "velvet brute" with a heavy-duty suspension that offered a "controlled but supple" ride. *Donald Farr Collection*

1965–66 Galaxie 427

By the beginning of the 1965 model year, Ford's Total Performance campaign was in full swing, including the continued availability of the 427 in the full-size Ford lineup, although limited to only the 425-horsepower Thunderbird High Performance V-8 with two four-barrel Holley carburetors. Also in full swing was the trend toward smaller and lighter cars. Pontiac lit the fire with the midsize 1964 GTO with a 389 and available Tri-Power induction.

Though big Galaxies had dominated Ford's performance landscape from 1961 to 1963, intermediates and sporty compacts started taking over in 1964 with the availability of the 289 High Performance in the Fairlane and Mustang. For the big-body Fords, emphasis began a shift from performance to luxury. The sporty 1965 500/XL, identified as "bucket-seat luxury," was joined by the 500 LTD, an upscale model designed for "elegance and dignified taste." The race-inspired, solid-lifter Thunderbird High Performance 427 was available in both.

After several years of mostly minor styling revisions to the big Ford's long, sleek body, the 1965 models were described as "new from road to roof." The overhaul started underneath with a stronger frame supported by a new suspension engineered to provide a smoother, quieter ride. New sheet metal all around brought the Galaxie into the swingin' sixties with a squared-off front end that incorporated stacked vertical

The Total Performance 1965 Ford Galaxie 500/XL 2-Door Hardtop

The velvet brute

Ford's still hanging tough on one rule: If you get a Big Ford with 425 horsepower you get it with heavy-duty suspension. Period.

For '65 we haven't changed the rule. Just the suspension. *And* any ideas you've had about how a firm-handling car couldn't be plushy.

This one is a paradox. Up front there's big muscle—427 cubic inches, two four-barrels, cross-bolted mains, 6000 rpm and 11.1-to-1 compression—the portrait of Brute Force. And when you get it rolling, don't look for a Brute Force ride. It's firm—but it's velvety firm. Controlled but supple.

It got that way because we started with Ford's new four-coil springing, locator-linkage rear suspension and "recessive" front wheels—the best expression yet of the soft-but-stable idea. All we had to do was pump in enough shock-dampening and spring-stiffness to match the 427's extra potential. Most of the velvet remained.

It sounds pretty simple but the *result* is pretty sophisticated. Matter of fact, there just isn't anything else like this Velvet Brute around. Try it—and if you still feel a red-hot performance car ought to ride like a dragster, even that needn't keep you out of an XL. Just run the tires up to 70 pounds and hang on.

Best year yet to go Ford!
Test Drive Total Performance '65

headlights, along with new hexagonal taillights at the rear, a major revision from the previous "tail-burner" round lenses. The new big Ford was offered in six different series, from the least-expensive Custom to the top-of-the-line LTD.

With the proliferation of lighter, nimbler cars like the Mustang, along with the Galaxie's new reputation for luxury, Ford sold only a handful of 1965 big Fords with the 427. Ford did not maintain production records prior to 1967, so the best estimate puts the number at 327, all with the R-code, dual-quad 427 with 425 horsepower. At mid-1965, the 427 was upgraded with a new block casting that fed main bearing oil from the side, an innovation that increased the engine's ruggedness for racing. The new engine, known as the side-oiler, was installed in a few 1965 Galaxies.

As with previous years, 1965 Fords with the 427 were equipped with mandatory heavy-duty components, including springs, shocks, and brakes, along with blackwall tubeless tires on 15-inch wheels, which became standard for full-size Fords in 1965. The rear axle was beefed with higher-capacity wheel bearings, larger axles, and a four-pinion differential. Belying the 1965 Galaxie's luxury image, the 427 was not available with power steering or brakes, air conditioning, or automatic transmission, with shifting restricted to the four-speed manual.

Despite its limited availability, *Car Life* acquired a 427-powered 1965 Galaxie 500/XL for a road test. "The biggest-engined Ford is not a car for the effete," it reported. "Indeed, it's tailored to the tastes of the knowledgeable and muscular." When describing the 427, the editors said,

> "When one pokes his toes into the carburetors, he must be prepared for the resultant action. If he shoves too far down on the throttle, his acceleration times soar in a cloud of tire smoke. If he's too timid, the engine tends to boggle at low rpm and the car sort of limps off the starting line. If he holds his toes just right, catches 3,500–4,000 rpm on the tachometer, and feathers the clutch so that power is fed gently to the rear wheels, he comes scrabbling off the starting line like a Super Stock champion."

Car Life's drag-strip efforts with the 4,400-pound 427 Galaxie resulted in a 14.9-second quarter mile at 97 miles per hour, a respectable clocking that could have been improved considerably, the reviewer said, with slicks, less restrictive mufflers, and more spark advance. In closing, the review questioned the car's warning light instrumentation and lack of power accessories, something that could have transformed the 427-powered 500/XL into an "Executive's Hot Rod."

1966–67 Galaxie 7-Litre

A new engine displacement and name debuted for the 1966 full-size Fords, which were essentially the same as 1965 but with a restyled grille and "Coke-bottle" rear quarters. Offered alongside the 427s, including the 425-horsepower dual-quad and the return of the 410-horse four-barrel, the 428-cubic-inch powerplant was yet another derivative of the FE, achieved by a combination of a 4.13-inch bore and 3.98-inch stroke. The 427's 4.23-inch bore stretched the limits of the FE block, so the 428 was also more cost-effective because it could be produced on a regular engine assembly line, not hand-built like the 427. For 1966, the new 345-horsepower 428 was designated as the "Thunderbird 7-Litre," although the name also crossed over to the 427, listed as an extra-cost option for the new 7-Litre model.

opposite top: The 1966 7-Litre Galaxie was designed to showcase Ford's new 428-cubic-inch FE big-block. *Mike Mueller*

opposite bottom left: At 345-horsepower, the 428 wasn't a brute like the 427, which was also available for the 7-Litre Galaxie. However, the 428 offered smoother power and could handle accessories such as power steering and air conditioning. *Donald Farr*

opposite bottom right: As an option for the XL, the 7-Litre was automatically equipped with bucket seats and console, especially fitting for a four-speed convertible like this one. *Tom Shaw*

FORD

7 LITRE

Although the 428 was available in all 1966 full-size Fords, including the station wagon, Ford elected to create a new Galaxie 500 7-Litre series to showcase the new big-block. The 7-Litre two-door hardtop or convertible came standard with the 428, XL bucket seats with console, woodgrain steering wheel, front disc brakes, stamped-steel full wheel covers, dual exhaust with resonators, nine-inch rear end, and 7 Litre emblems. Available with four-speed or automatic, the 7-Litre bridged the gap between the race-inspired, solid-lifter 427 and the Galaxie's emerging elegance and luxury image. Compared to the finicky, noisy 427, the hydraulic-lifter 428 was powerful, smooth, and fully capable of driving power steering, air conditioning, and other creature comforts.

Car Life tested a new Galaxie 500/XL two-door hardtop with automatic transmission and proclaimed it "a six-quart package of performance." However, the editors were not particularly impressed with the 428 as a performance engine, blaming the 7-Litre's mild camshaft and small four-barrel carb after clocking a disappointing 16.4-second quarter-mile time, some 1½-seconds slower than the previous year's Galaxie with a dual-quad 427.

Although not a barnstormer like the 427, the 7-Litre succeeded where it counted—in sales. As a "luxury car with a high degree of performance or a performing car with a high degree of luxury," per Ford's description, the well-rounded 7-Litre found plenty of buyers. According to the 7-Litre Registry, Ford sold more than eleven thousand 7-Litre Galaxies for 1966—more than all 427 Galaxies combined over the previous three years—the huge majority as hardtops with 428 and automatic transmission. Fewer than forty were equipped with the 427.

7 LITRE
1966 Ford 7-Litre Convertible
Ford 7-Litre...either the quickest quiet car or the quietest quick car
Well, once again we've invented a new kind of car. It's not a competition car (that's why the overbore to 7 litres/428 cubic inches.) But it turns on like a competition car (after all, 462 pounds/feet of torque!) What it is is lightning without thunder. It *moves*—but it moves like mist over a millpond, smoothly, quietly, effortlessly!
It *stops*, too! Power disc brakes up front are standard. So are bucket seats. The V-8 comes in just one size, with a 4-barrel carburetor and the beefy bottom end that is the heritage of Ford's tremendous competition program. But the lifters are hydraulic for silence' sake and even the dual exhausts are very discreet. You get your choice of convertible or two-door hardtop, four-on-the-floor or Cruise-O-Matic . . . and just about any other added pleasure Ford makes, including air conditioning.
You'll have to decide whether it's a cool hot car or a hot cool car. But one thing you're bound to decide—there just isn't anything else like it!

AMERICA'S
TOTAL PERFORMANCE CARS
FORD
MUSTANG • FALCON • FAIRLANE
FORD • THUNDERBIRD

Ford's advertising for the 1966 7-Litre Galaxie emphasized that the new 428 wasn't a competition engine like the 427. *Donald Farr Collection*

Ford changed up the big Ford again for 1967, adding 3 inches in length and an aluminum grille reshaped into a pointy nose. The Galaxie name was dropped for the top-of-the-line models—they were known simply as the Ford XL and LTD. Ford also demoted the 7-Litre to a $515 XL option, described in the sales brochure as the 7-Litre Sports Package with the 345-horse Thunderbird 428, Cruise-O-Matic transmission, low-restriction dual exhaust, power front disc brakes, Wide Oval tires, special suspension, and woodgrain steering wheel. While the 428 remained an option in other 1967 full-size Fords, only cars with the option package received the 7 Litre emblems. With less marketing emphasis, Ford sold only 1,068 Ford XLs with the 7-Litre Sports Package, including 12 with the still-optional 427, 2 with the 410-horse four-barrel, and 10 with the 425-horse dual-quad engine.

The 427 was also available in other 1967 full-size Fords, but only eighty-nine were sold, seventy with the R-code and nineteen with the single four-barrel W-code.

When the 1968 Fords were introduced, the 427 was missing from the sales brochure's list of available engines. The glory days of the 427 Galaxie had come to an end.

1966–67 Mercury S-55

Touted as "moving ahead in the Lincoln Continental tradition," Mercury left little doubt that its 1966 full-size lineup of Park Lane, Montclair, and Monterey was more about luxury than performance. Growing 2 inches in length and weighing in at well over two tons, the 1966 Mercury conceded its muscle car image to Ford. While the 427 remained available for the Galaxie, the solid-lifter FE was nowhere to be found on the big Merc's option list. However, sporting enthusiasts still had one vestige of Mercury's performance past—the S-55.

Like the Galaxie 7-Litre, the "sleek, smooth, and sassy" 1966 Mercury S-55 was offered as a series of its own for the two-door hardtop and convertible with bucket seats, console, floor shifter, full wheel covers, and S-55 ornamentation. It was available only with the 345-horsepower Super Marauder 428 with dual exhaust and four-speed, with automatic as an option. For the "man around town" looking for excitement in a big package, the S-55 provided plenty of "get up and go" with a combination of luxury appointments and sporty interior. Mercury sold 2,916 S-55 hardtops and 669 convertibles for 1966.

The 1966 S-55 was Mercury's counterpart to the Ford 7-Litre Galaxie, a big luxury two-door hardtop or convertible with bucket seats, console, and a 428. *Archives / TEN: The Enthusiast Network Magazine, LLC*

above: When optioned with bucket seats, the X-100's Select-Shift automatic was controlled by a console-mount floor shifter. Other than fuel gauge and clock, the X-100 contained limited instrumentation; the owner of this one has added aftermarket gauges.

opposite: The matte black rear end treatment was a signature graphic for the 1969 X-100. However, half of the X-100 buyers chose to delete it as a cost-reduction option. *Cam Hutchins*

For 1967, the big Mercury sports car image succumbed to luxury models such as the Marquis but meekly continued with the S-55 Sports Package, an option for the convertible and two-door hardtop, which gained a new "swept-back" roofline. Under the long hood, 1967's Super Marauder 428 received a dress-up kit with chrome appointments for the rocker arm and air cleaner. Full wheel covers came with spinners, and the standard equipment listing was updated with power steering, heavy-duty battery, and "luxury car" insulation. The S-55 Sports Package was phased out during the year, resulting in sales of only 570 hardtops and 145 convertibles.

1969–70 Marauder X-100

By 1969, America's muscle car landscape was enveloped by cars such as the intermediate-size Fairlane Cobra and pony car Mach 1. The full-size models that had reigned supreme just a few years earlier had succumbed to "big boat" luxury status. But that didn't stop Mercury from giving it one more shot with the 1969–70 Marauder X-100.

Based on the huge Marquis, the Marauder was a sportier two-door version with a Lincoln-like front end and sweeping tunnel-window roofline. In standard

Marauder
Collector
B13-709

form, the Marauder came with the 265-horsepower 390, but the X-100 was powered exclusively by the 360-horsepower 429 four-barrel with Select-Shift Cruise-O-Matic, either column shift or supplied with floor shift when optioned with bucket seats and console. Priced at $700, the X-100 package also came with styled aluminum wheels with Polyglas tires, fender skirts, rim-blow steering wheel, electric clock, woodgrain interior trim, and small "X-100" chrome emblems. However, the most distinguishable feature was the "Sports-Tone" matte black finish on the rear deck.

Car Life's road test of the 1969 Marauder X-100 netted a 15.16 quarter mile—"almost a supercar," it said—but reviewers discovered that the big Merc was best on the open road: "The 429 engine is strong at any speed and the gearing is well chosen. All the power the driver can use is available whenever he wants."

Car & Driver praised the X-100's optional competition suspension: "The shock absorber control is such that you hit a bump and feel it just once rather than going through the diminishing oscillations for half a block. You know there's a road under you because you can feel it and that's very reassuring."

Even at its hefty $4,500 sticker price, with options pushing the total to well over $5,000, the X-100 sold well, with 5,635 produced for 1969. The package continued mostly unchanged into 1970 with an additional 2,646 sold.

Several auto magazines put the big and heavy Mercury X-100 through their standard testing procedures. In spite of its nearly 4,600-pound weight, the 429-power Merc ran low 15-second ETs. *Archives / TEN: The Enthusiast Network Magazine, LLC*

BY THE NUMBERS:

1961 GALAXIE STARLINER

ENGINE: Thunderbird Super 390

CARBURETION: Holley four-barrel

HORSEPOWER: 375 at 6,000 rpm

TORQUE: 427 at 3,400 rpm

TRANSMISSION: Three-speed overdrive

REAR AXLE RATIO: 4.10

WEIGHT: 3,723 lbs (dry)

HORSEPOWER to weight: 9.92

QUARTER MILE: N/A

Archives / TEN: The Enthusiast Network Magazine, LLC

1962 MONTEREY S-55 CONVERTIBLE

ENGINE: Super Marauder 406

CARBURETION: Three Holley two-barrels

HORSEPOWER: 405 at 5,800 rpm

TORQUE: 448 at 3,500 rpm

TRANSMISSION: Four-speed

REAR AXLE RATIO: 3.56:1

WEIGHT: 4,590 lbs (test)

HORSEPOWER TO WEIGHT: 11.33

QUARTER MILE: 16.1 at 85 mph (*Car Life*, September 1962)

1962 GALAXIE TWO-DOOR SEDAN

ENGINE: 406 Super High Performance

CARBURETION: Three Holley two-barrels

HORSEPOWER: 405 at 5,800 rpm

TORQUE: 448 at 3,500 rpm

TRANSMISSION: Four-speed

REAR AXLE RATIO: 3.56:1

WEIGHT: 4,210 lbs (test)

HORSEPOWER TO WEIGHT: 10.39

QUARTER MILE: 15.3 at 93 mph (*Car Life*, March 1962)

BY THE NUMBERS:

1964 GALAXIE 500 TWO-DOOR HARDTOP

ENGINE: 427 Thunderbird 8V

CARBURETION: Two Holley four-barrels

HORSEPOWER: 425 at 6,000 rpm

TORQUE: 480 at 3,700 rpm

TRANSMISSION: Four-speed

REAR AXLE RATIO: 4.11:1

WEIGHT: 4,000 lbs (approximate)

HORSEPOWER TO WEIGHT: 9.41

QUARTER MILE: 15.4 at 95 mph (*Motor Trend*, February 1964)

1965 GALAXIE 500/XL

ENGINE: Thunderbird High Performance 427

CARBURETION: Two Holley four-barrels

HORSEPOWER: 425 at 6,000 rpm

TORQUE: 480 at 3,700 rpm

TRANSMISSION: Four-speed

REAR AXLE RATIO: 3.50:1

WEIGHT: 4,426 lbs (test)

HORSEPOWER TO WEIGHT: 10.41

QUARTER MILE: 14.9 at 97 mph (*Car Life*, February 1965)

1963½ S-55 TWO-DOOR HARDTOP

ENGINE: Super Marauder 427

CARBURETION: Two Holley four-barrels

HORSEPOWER: 425 at 6,000 rpm

TORQUE: 480 at 3,700 rpm

TRANSMISSION: Four-speed

REAR AXLE RATIO: 4.11:1

WEIGHT: 4,485 lbs (test)

HORSEPOWER TO WEIGHT: 10.55

QUARTER MILE: 15.1 at 87 mph (*Car Life*, April 1963)

1966 GALAXIE 7-LITRE

ENGINE: Thunderbird 428

CARBURETION: Four-barrel

HORSEPOWER: 345 at 4,600 rpm

TORQUE: 462 at 2,800 rpm

TRANSMISSION: Automatic

REAR AXLE RATIO: 3.25:1
Weight: 4,490 lbs (test)

HORSEPOWER TO WEIGHT: 13.01

QUARTER MILE: 16.4 at 89 mph (*Car Life*, January 1966)

1969 MARAUDER X-100

ENGINE: 429 4V

CARBURETION: Four-barrel

HORSEPOWER: 360 at 4,600 rpm

TORQUE: 480 at 2,800 rpm

TRANSMISSION: Select-Shift Cruise-O-Matic

REAR AXLE RATIO: 2.80:1

WEIGHT: 4,580 lbs (test)

HORSEPOWER TO WEIGHT: 12.72

QUARTER MILE: 15.17 at 92.3 mph (*Car Life*, April 1969)

1966 MERCURY S-55

ENGINE: Super Marauder 428

CARBURETION: Single four-barrel

HORSEPOWER: 345 at 4,600 rpm

TORQUE: 462 at 2,800 rpm

TRANSMISSION: Four-speed or automatic

REAR AXLE RATIO: N/A

WEIGHT: 4,260 lbs (test)

HORSEPOWER to weight: 12.34

QUARTER MILE: 16.9 at 84 mph (*Motor Trend*, August 1966)

Archives / TEN: The Enthusiast Network Magazine, LLC

TEXAS
CH·5921

CHAPTER TWO

COMPACTS & INTERMEDIATES

1963–1967

By 1964, Ford's Total Performance pendulum was swinging from big-engined, big-bodied Galaxies to smaller, lighter compacts and intermediates powered by the new, lightweight Windsor small-block engine. Even the economy-based Falcon and Comet got into the act with peppy Sprint and Cyclone models. But the heart of the emerging muscle car market would be based on the intermediates, a trend fueled by Pontiac's GTO, introduced in 1964 as a Tempest with a 389-cubic-inch engine, available Tri-Power, and an image name.

Strangely, the Ford and Mercury intermediates evolved from both ends of the size spectrum. In 1962, Ford's Fairlane was downsized from its original 1955–61 big-body form, and for 1966 Mercury's Comet shed its compact Falcon roots by growing into an attractive midsize based on the Fairlane. Both entered 1964 with available 289 power. By 1967, they would rip up the streets and strips with a 427 under the hood.

At midyear 1963, the Fairlane gained respect with the addition of the 271-horsepower 289 High Performance to the engine lineup. *Jerry Heasley*

The Fairlane and Comet entered the Total Performance era as economy cars, but by 1967 both were packing 427 power. *Mike Mueller*

Rick Adis was an admitted "street racer for money" when he ordered a 427-powered 1967 Fairlane 500XL to replace his 1958 four-door Fairlane with a transplanted 406. Rick recalled, "I had to order the Fairlane four times because Ford kept saying the 427 wasn't available to the public." The fourth time was the charm, with Rick taking delivery in March 1967 after shelling out $1,700 extra for the 427. When Rick discovered that the factory distributor limited engine revs to 5,000 rpm, he replaced it with a 427 Galaxie distributor for 7,000 rpm shifts and resulting 12-second quarter-mile ETs, although Rick said he rarely visited the drag strip because he didn't want to draw attention to the car's capability. Covertly, he even replaced the 427 fender emblems with 390 versions. "I won a lot of money from other street racers," admitted Rick, who still owns his 427 Fairlane.

1963½–65 Fairlane 289 Hi-Po

The Fairlane name first appeared in 1955 as a full-size Ford, where it remained through 1961. For 1962, however, the Fairlane designation shifted to a new Ford intermediate positioned between the compact Falcon and full-size Galaxie. For the first year and half, the Fairlane was a utilitarian passenger vehicle with six-cylinder or small V-8, even for the Sport Coupe two-door with bucket seats and spinner wheel covers. That changed in early 1963 as Ford's Total Performance campaign launched in earnest.

In March 1963, around the same time as the introduction of the Falcon Sprint and the 427 for the new fastback Galaxie, Ford added a "Challenger" 289 High Performance V-8 to the Fairlane's engine options. The new small-block, designated by a "K" engine code in the VIN, made its 271 horsepower at 6,000 rpm thanks to solid-lifters, 10.5:1 compression, dual-point distributor, higher-flow exhaust manifolds, and

595 cfm four-barrel carburetor. For $424, the 289 Hi-Po was available in all 1963 Fairlanes except the station wagon and only with a manual transmission. A package deal, the Hi-Po also added heavy-duty springs and shocks, a higher-capacity radiator, sturdier transmission output shaft, heavy-duty clutch, and 9-inch rear axle. Strangely, the 1963½ Hi-Po Fairlanes were equipped with a single exhaust system, although with larger 2½-inch pipes.

The 1964 Fairlanes were in dealer showrooms before *Car Life* magazine could obtain one with a Hi-Po for testing. With only minor sheetmetal and body trim updates, the 1964 Hi-Po Fairlane was mechanically similar to the earlier 1963s. The 3,100-pound hardtop tested by Roger Huntington registered a 16.18-second quarter mile at 87.12 miles per hour. "This new 1964 high-performance Fairlane 289 would seem to be an interesting car," Huntington reported. "The ride is good, the handling is decent, the gas mileage appeared to be over 15 mpg in normal driving, and the performance is very good for the price class."

In the spring of 1964, the Hi-Po Fairlane sprouted much-needed dual exhaust. Even better, the new system was manufactured by Arvin Industries, a company recognized for its sound attenuation. Called Arvinode, the setup included a pair of glasspack-type

When Pontiac dropped a 389 into the intermediate Tempest and called it the GTO, the muscle car tide began turning from big bodies to lighter, more nimble midsizes. *Archives / TEN: The Enthusiast Network Magazine, LLC*

HIGH PERFORMANCE
289

top: Ford introduced the 289 High Performance small-block for the Fairlane at midyear 1963. With solid-lifters, a four-barrel carb, and a dual-point distributor, it made 271 horsepower at a high-revving 6,000 rpm. *Donald Farr Collection*

oppposite top: A four-speed stick and bench seat were common for early Hi-Po Fairlanes. *Jerry Heasley*

oppposite bottom left: A "High Performance" plate behind the standard 289 emblem was the only external identification for a Hi-Po 1963 Fairlane. *Jerry Heasley*

oppposite bottom right: From "mild to wild," all the way up to the 289 High Performance, was how Ford promoted the 1964 Fairlane's V-8 engines. *Donald Farr Collection*

inline mufflers with "wave-tuned" resonators. It produced a rumble, one that pleased performance enthusiasts but annoyed buyers who preferred a quieter ride from their midsize Fairlane.

At the same time, Ford added the C4 automatic to the option list, promoting it heavily in a series of full-page advertisements.

The 289 High Performance option continued for the 1965 Fairlane, which received a "wider, more massive" facelift that *Motor Trend* also described as "bread-and-butter." The Arvinode dual exhaust and optional Cruise-O-Matic transmission continued, and the Sports Coupe was available with an optional console selector similar to the Galaxie 500/XL.

No Ford production records were maintained prior to 1967, but Bob Mannel from the Fairlane Club of America estimated that around 1,350 Hi-Po Fairlanes were sold for 1963's half-year production, followed by some 3,600 during 1964. Sales fell to less than 800 for 1965 as Ford performance enthusiasts instead chose the sportier Mustang.

289 High Performance

In 1961, Ford introduced a new Windsor small-block V-8 for the 1962 Fairlane and Mercury Meteor. Compact and lightweight thanks to its thin-wall casting, the 221-cubic-inch engine was ideal for the increasingly popular midsize Fords of the early 1960s. At midyear 1962, the bore was increased to displace 260 cubic inches, big and powerful enough to replace the Y-block as the base V-8 for 1963's full-size Fords. In April 1963, the Windsor was enlarged again to 289 cubic inches.

The 289 was ideal for factory power enhancement. The High Performance version arrived midyear 1963 for the Fairlane. Rated at 271 horsepower at 6,000 rpm, the Hi-Po was prepped for high-revving horsepower with a solid-lifter camshaft, dual-point distributor, 595 cfm four-barrel carburetor, header-style exhaust manifolds, and cylinder heads with smaller combustion chambers, cast spring cups, and screw-in rocker arm studs. Inside the short-block, the engine sustained its high rpm power with thicker main bearing caps, high-nodularity crankshaft, and larger rod bolts, which mandated revised crankshaft counterweighting and a new balancer. Hi-Po 289s were also equipped with a high-revving water pump and larger pulley for the generator (1963–64) or alternator (1965 and later).

Initially used in the Fairlane, the 289 High Performance became an option for the Mustang shortly after its April 1964 introduction. Hi-Po Fairlane and Mustang VINs had a "K" for their engine code, thus the 289 High Performance became known as the "K code." For Shelby's 289 Cobras and 1965–67 GT350s, the 289 High Performance was upgraded to Cobra status with an aluminum intake, Holley four-barrel, and headers for 306 horsepower.

From 1963 to 1967, the 289 High Performance was the top-performing Ford small-block. It was replaced by a four-barrel 302 in 1968.

Put away the boring bar–
we've done it for you!
Now we've scooped out Fairlane's V-8 to
289 cubes . . . 271 h.p.! This, friend, is a real
stormer! Solid lifters, 4-barrel carb, the whole
bit. Can you think of better news for the guy
who wants solid, off-the-line punch from a
gem-size power plant? Tie this savage little
winder to a four-speed floor shift, tuck it into
Fairlane's no-fat body shell, and you've got a
going-handling combo that's mighty hard to
beat...and we mean that both ways! The factory overbore is better than doing it yourself
. . . you know the cores are in the right place,
and the bottom end is tested to take the
kind of rpm this V-8 churns out! So put away
the boring bar and check your friendly Ford
Dealer; he'll show you what 289 cubes can do.
America's liveliest,
most care-free cars!
APRIL, 1963
13

FORD
MOTOR COMPANY

above: When Carroll Shelby sank his fangs into the 289 High Performance, it became the 306-horsepower 289 Cobra with Holley four-barrel, aluminum intake, and Tri-Y headers. It was used for the 1962–65 Cobras and 1965–67 GT350 Mustangs. *Archives / TEN: The Enthusiast Network Magazine, LLC*

left: In 1963, Ford described the 289 High Performance as a "factory overbore," giving performance drivers more cubic inches for a "savage little winder." *Donald Farr Collection*

1963½–65 Falcon Sprint

If ever a two-barrel Ford deserved muscle car status, it was the Falcon Sprint, which joined the Hi-Po Fairlane and 427 Galaxie at midyear 1963 as part of Ford's Total Performance campaign.

Weighing less than 3,000 pounds as a two-door hardtop, the Sprint's standard 164-horsepower, 260-cubic-inch V-8 with four-speed transmission made the little Falcon a "real goer," per Ford's sales copy. Offered as both a hardtop and convertible, the Sprint started with Futura equipment and then added bucket seats, spinner wire wheel covers, chrome engine dress-up, simulated woodgrain steering wheel, and a tachometer mounted to the top of the dash. For a gutsy sound, the Sprint was also equipped with an open-element "power hum" air cleaner and "throaty" muffler. When introduced in 1960, the Falcon was envisioned as an economy compact, not a performance car, so the Sprint package also added a stiffer suspension and chassis, larger 10-inch drum brakes, and five-lug wheels to handle the V-8's power and weight.

According to the Falcon Club of America, Ford sold more than 15,000 Falcon Sprints during its abbreviated half-year 1963 run, 10,479 hardtops and 4,602 convertibles.

The Falcon received its first facelift for 1964, abandoning 1960–63's soft and round sheet metal for a sharp-edged, boxy shape. The Sprint continued with its 164-horse 260, but in January 1964 it became "Sprint flair at a new lower Sprint price" with the bucket seats, console, and tachometer moving to the option list. For 1964, all Falcons were available with the Challenger 260-cubic-inch V-8. However, the V-8 remained the standard powerplant in the Sprint, which also added chrome engine dress-up, open-element air cleaner, and special muffler. Available Sprint transmissions for 1964 included the three- and four-speed manuals and automatic.

bottom right: The midyear 1963 Sprint package put some pizzazz into the Falcon. *Dick Harrington*

below: The Falcon Sprint was part of Ford's 1963½ "Lively Ones" advertising along with the Galaxie Sports Hardtop and Fairlane Sport Coupe. *Donald Farr Collection*

REPORT FROM MONACO
Ford premieres the Liveliest of the Lively Ones-
new Command Performance Cars for 1963½
A new Royal Family of Fords has just
made its bow before the car-wise audi-
ence that assembles each year for
Europe's most famous road rally. The
verdict: Vive la Ford! That regal roofline
in the foreground (looks like a convertible but isn't) crowns the new Super
Torque Ford Sports Hardtop. At left
background: new Fairlane Sports Coupe
offers a choice of two V-8s. At right:
the hot, new Falcon Hardtop that intro-
duces scatback styling to the compact
field. American première at
your Ford Dealer's!
America's liveliest,
most care-free cars
FORD

Car Life tested a 1964 Sprint convertible, and, although the heavier ragtop didn't set the muscle car world on fire with its 18-second quarter-mile time, the road testers were pleasantly surprised by the Sprint's demeanor, describing it as "So thoroughly delightful, so eminently practical, and so purposefully sporting that most of our staff members were constantly seeking errands to run in order to borrow it."

Although the Mustang stole the Falcon Sprint's thunder after April, Ford sold nearly 17,500 Sprints for 1964. Not surprisingly, most were ordered with extra-cost bucket seats and console.

Externally, the 1965 Falcon Sprint appeared very similar to the 1964. The big change came under the hood by upgrading to the larger 289. Still with two-barrel carburetion, the power jumped to 200, a significant 36-horsepower improvement over 1964. Overshadowed by the Mustang, the Sprint was downplayed in promotional material, which showed it as a package for convertibles and hardtops with reduced content, primarily the standard 289 engine, bucket seats, and "289 Sprint" fender emblems. The console was standard for the Sprint convertible but optional in the hardtop. Reportedly, the 289 High Performance was not offered in the 1965 Falcon except for Canada, where few were sold.

Not surprisingly, the new Mustang sucked the air out of 1965 Sprint sales with production falling to just over 3,000, mostly hardtops with 2,806, sold plus 300 convertibles.

For 1964–65, the Falcon adopted a boxier shape, still with wire wheel covers for the Sprint option. This is a 1965 model with the 289. *Barbara McCray*

1964–65 Cyclone

While the 1964 Falcon Sprint was limited to a 164-horse 260, Mercury upped the ante for the Comet with the availability of the "Super 289" for the new Cyclone. Confusingly, it carried a "K" engine code in the VIN, but the Mercury version was not the solid-lifter, 271-horse 289 High Performance as found in the K-code Fairlane. Instead, the Cyclone's engine was a four-barrel, hydraulic-lifter 289 with 210 horsepower, a peppy package nonetheless for the lightweight Comet.

Based on the restyled 1964 Comet, the Cyclone model replaced the previous S-22 to provide the Mercury compact with a more exciting image. Available only for the two-door hardtop, the Cyclone package included bucket seats, console, dash-mount tachometer, three-spoke "rally" steering wheel, and chrome valve covers, air cleaner lid, and dipstick. "Race-like" chrome wheel covers were the most distinctive feature, as found only on the Cyclone. A three-speed manual was the standard transmission, with the Merc-O-Matic and floor-shift four-speed available optionally. Like the Falcon Sprint, the 1964 Cyclone was bolstered for V-8 power with heavier-duty suspension, larger drum brakes, and five-lug wheels.

For 1965, the Cyclone climbed another rung up the performance ladder when its standard Super 289 small-block gained another 15 horsepower, up to 225 like the Mustang's four-barrel 289 (and unlike the Falcon Sprint's two-barrel 289). New options included a performance handling package with quicker ratio steering, higher rate springs, and larger sway bar, plus a Rally Pac three-gauge cluster on the dash that replaced the previous tachometer with a smaller tach, vacuum gauge, and elapsed time clock. The 289 High Performance was a little-known option by special order; reportedly, fewer than one hundred were built.

Motor Trend clocked a 16.2-second quarter mile for the 289-powered 1964 Cyclone. *Archives / TEN: The Enthusiast Network Magazine, LLC*

Comet
463 ME
1965 OHIO

top: While the Falcon Sprint got a two-barrel 260, the Cyclone's Super Cyclone 289 generated 210 horsepower with four-barrel carburetion. *Donald Farr*

bottom: The Cyclone package included bucket seats, console, and three-spoke steering wheels. This 1965 Cyclone is equipped with the optional Rally Pac three-pod gauge cluster on the dash that replaced the Cyclone's standard dash-mount tachometer. *Mike Mueller*

opposite: Model year 1965 would be the final year for the Cyclone on a compact platform. Chrome wheel covers were an identifying feature for the 1964–65 Cyclones. *Mike Mueller*

Motor Trend tested a 1965 Cyclone with four-speed and optional 3.50 gearing, describing it as "one of the most tasteful renderings to come out of the Mercury styling studios in many a moon." Noting that the transmission was a new Ford-built four-speed instead of the older Borg-Warner design and that the rear end could have used a locking differential, the *Motor Trend* testers coaxed a 16.2-second, 82-mile-per-hour quarter mile clocking out of the 3,160-pound Cyclone. As the editors pointed out, the Cyclone provided Mercury with "a little more ammunition to stave off the Mustang attack."

1966–67 Fairlane GT/GTA 390

Using *Car Life's* formula of 12 pounds per horsepower as a measuring stick for a bona fide "supercar," the 1966 Fairlane GT/GTA easily made the cut. Attempting to counter-punch Pontiac's GTO, Ford slapped a similar image name on its freshly restyled intermediate, packed it with a standard 335-horsepower 390 Thunderbird Special, and sent it out onto the shark-infested streets to battle not only GTOs but also General Motors' latest crop of supercars, including the Skylark Gran Sport, Chevelle Super Sport 396, and Cutlass 4-4-2.

Not only was the 1966 Fairlane GT/GTA available only with the 390 Thunderbird Special, it was the only way to get the 335-horse engine in a Fairlane. Offered for convertibles and hardtops, the GT/GTA package also added stiffer springs, larger front sway bar, bucket seats with console and floor shifter, blackout grille, side striping, and nonfunctional hood ornamentation. GTs got the three- or four-speed manual, but when ordered with Ford's new C6 Cruise-O-Matic transmission, the designation became GTA for "GT Automatic."

Ford had good reason to highlight its new "Sports Shift" Cruise-O-Matic. For the first time, drivers could select the "D" shifter position for fully automatic operation, or they could upshift and downshift for themselves as a cross between fully automatic and manual.

Car Life pointed out that the 1966 Fairlane GTA was not the ideal answer to the GTO, but it was a "good start." In spite of its snappy chrome dress-up kit, the 390 wasn't up to the task of hanging with 389 Pontiacs, 401 Oldsmobiles, and 396 Chevys. For the GT/GTA, the 390 was equipped with a 600 cfm Holley four-barrel on a revised intake manifold and slightly hotter camshaft for 20 more horsepower than the standard four-barrel 390. The cylinder heads were also improved but not enough to unleash the 390's potential, leading *Car Life* to report, "It chokes up, flattens out, falls off so badly beyond 4,400 rpm that real storming stripsmanship is out of the question." The 3,880-pound test Fairlane GTA ran a 15.4-second quarter at 87 miles per hour, with the editors noting that available speed equipment—six-barrel intake, reworked heads,

HOW TO COOK A TIGER
Take one part 335 HP V-8. Chrome plate the rocker covers, oil filler cap, radiator cap, air cleaner cover and dip stick.
Blend high lift cam; bigger carburetor.
Mix in the new 2-way, 3-speed GTA Sport Shift that you can use either manually or let shift itself.
Place the new shift selector between great bucket seats.
Now put on competition type springs and shocks.
Add a heavy-duty stabilizer bar.
Place over low profile 7.75 nylon whitewalls.
Touch off with distinctive GTA medallion and contrasting racing stripe.
Cover with hardtop or 5-ply vinyl convertible top with glass rear window. Serve in any of 15 colors.
This is the new Fairlane GTA. An original Ford recipe that may be tasted at your Ford Dealers . . . Remember--it's a very hot dish!
FAIRLANE
GTA
A PRODUCT OF
Ford

Ford's "recipe" for cooking a tiger—a reference to the GTO's tiger theme—may have been wishful thinking. The GTO's optional 400 HO boasted 25 more horsepower than the 1967 Fairlane's 390. *Donald Farr Collection*

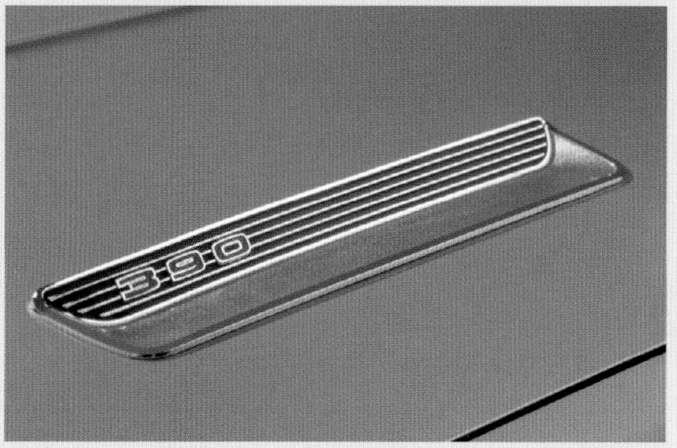

top: The 1966 Fairlane moved solidly into the 1960s with a complete styling overhaul, one that incorporated stacked headlights and the "Coke bottle" profile that was popular at the time. *Jerry Heasley*

above: The 1966 GT/GTA hood ornaments were nonfunctional, other than identifying the 390 under the hood. *Jerry Heasley*

left: When equipped with the new Sports Shift Cruise-O-Matic, the 1966 Fairlane GT became the GTA—for "GT Automatic." *Jerry Heasley*

Motor Trend's fifth-wheel acceleration test of a 1966 Fairlane resulted in a 15.2-second quarter-mile clocking. *Archives / TEN: The Enthusiast Network Magazine, LLC*

headers, slicks, and 3.50 gears—would put high 13s within reach of a good driver.

Motor Trend fared somewhat better in the quarter mile, lowering the GTA's quarter mile clocking to 15.2 seconds during a test session at Ford's ride and handling course. "Holding the brake, bringing the revs up to 1,500, releasing, and stabbing it produced slingshot-like acceleration every time," the writer said. "There was enough power available to keep the rear wheels smoking nearly the entire length of the rather tight handling course."

For 1967, the Fairlane went through the typical second-year styling updates, mainly a new front grille and revised GT/GTA hood ornamentation, now called "power domes" with integral turn signal indicators. The GT/GTA powertrain choices expanded to include the 289 Challenger and 390 Thunderbird V-8, both with two-barrel induction. The 390 Thunderbird Special was available optionally and only for the GT/GTA, but the rating dropped 15 horsepower to 320, although it was the same engine as 1966.

For 1967, Ford produced 15,196 Fairlane GT/GTAs with the 390 big-block, 6,303 as GTs and 8,893 as GTAs.

1966–67 Cyclone GT 390

For 1966, the Comet abandoned its compact Falcon heritage and became Mercury's cousin to Ford's midsize Fairlane, which had gone through a complete overhaul for 1966. Like the Fairlane, the larger engine compartment accommodated the 390 big-block to place the Comet Cyclone GT squarely in the supercar category. In one model year, the Cyclone's power vaulted from a 225-horsepower 289 in 1965 to a 335-horsepower 390 GT for 1966.

While the base Cyclone was powered by the 289, the Cyclone GT was the only way to get the 335-horsepower 390 with Holley four-barrel and hotter cam, as also found in the Fairlane GT/GTA. The $452 GT package was more than a big-block engine; it also added dual exhaust and Special Handling Package with 5½-x14-inch wheels. The manual three-speed was standard with optional four-speed and Sport Shift Merc-O-Matic—same as Ford's new Select Shift Cruise-O-Matic—all with floor shift. Externally, the Cyclone GT was identified by its side stripes, chrome wheel covers, and fiberglass hood with nonfunctional twin scoops. Mercury sold 15,970 Cyclone GTs in 1966—13,812 hardtops and 2,158 convertibles.

Car Life tested an automatic 1966 Cyclone GT on the quarter mile and came away with mixed emotions after running a 15.2-second elapsed time. "By strange admixture of mechanical alchemy, the 390 four-barrel has never been much of a top-end performer," the editors noted. "It develops plenty of usable torque in the lower reaches, and it pumps up more than enough horsepower for its nominal purpose. But as a performer, it just doesn't deliver."

Car & Driver clocked an impressive 13.98-second quarter mile at 103.9 miles per hour from its 1966 Cyclone GT, but its review was quick to point out—and was somewhat miffed—that the test car had been prepped by Bud Moore Engineering, Mercury's NASCAR race shop.

CYCLONE
GT
1966
LINCOLN-MERCURY DIVISION OF
Ford
Hottest new entry in the whole blazing GT world.
Performance fans! Here's your breakfast, lunch, dinner and midnight snack. The big, roomy, new Comet Cyclone GT. With a new 390 4-barrel V-8 roaring under its twin scoop hood, this Cyclone GT delivers go that can shove you right back into your bucket seat. Also included: console mounting for the transmission you choose (there's a 4-speed manual specially geared to be quick on the takeoff); heavy-duty, wide-rim wheels; engine dress-up kit; fade-resistant brakes. Plus high-rate front and rear springs, a big-diameter stabilizer bar, and HD front and rear shocks. Add the optional tach and this car's ready to rally! And because it's a new-generation Comet, there's special luxury here. You'll see it in the interior trim and feel it in the soft vinyls and carpets. See all the bigger, livelier new '66 Comets. Eleven of them cover the field from a rakish Comet 202 to the handsome Capris and Calientes and this blazing Cyclone GT. They're at your Mercury dealer's. Now.
the big, beautiful performance champion
Mercury Comet
GT

"This Cyclone GT delivers go that can shove you right back into your bucket seat," said the ad copy for the 1966 Cyclone GT. *Donald Farr Collection*

The midsize Mercury's top performance model continued for 1967 as the Cyclone GT Performance Group option, which contained the same equipment as 1966. As with the Fairlane GT/GTA, the four-barrel 390 horsepower was down-rated to 320. Sales dropped to 3,797, all hardtops except for 378 convertibles.

1966–67 Fairlane 427

For the 427 to compete in NHRA Stock and Super Stock drag racing, Ford needed to produce at least fifty 427-powered street Fairlanes. Thus the 427 Fairlane was born at mid-1966 as a limited production option.

Dropping the Medium Riser 427 into the midsize Fairlane created the ultimate muscle car—425 horsepower in a lightweight intermediate. The 427 Fairlanes were instantly recognized by their forward-scooped fiberglass hood, which rammed cooler outside air into the engine compartment while also reducing weight. The hood did not use hinges, springs, or a latch; it was simply retained by racing-style hood pins at the four corners.

With dual-quad 427 power under the hood, the rest of the two-door Fairlane was beefed for the task with a heavy-duty four-speed, 9-inch rear axle with thirty-one spline axles, 11.2-inch front disc brakes, free-flowing exhaust manifolds, an extra cooling package, Fairlane GT suspension, and a lightweight package that deleted seam sealer and sound deadener. Reportedly, Ford planned to build at least seventy 427 Fairlanes

To legalize the 427 for NHRA drag racing, Ford produced fifty-seven production 427 Fairlanes for 1966, all white with fiberglass hoods. *Mike Mueller*

Hot Rod's editors were fascinated by the 1966 Fairlane 427's fiberglass hood, which was retained by four hood pins for "instant status when you calmly step out of your Fairlane at the gas station and gingerly pull the pins, lift off the hood, and check the oil." *Archives / TEN: The Enthusiast Network Magazine, LLC*

for 1966, but the late introduction combined with delayed casting of the exhaust manifolds limited production to only fifty-seven, all white two-door hardtops with dog-dish hubcaps on steel wheels. Many, if not most, were quickly converted into Stock or Super Stock drag cars by their owners.

To evaluate the new 427 Fairlane, *Hot Rod* sent writer Eric Dahlquist to Dearborn where Ford provided a sneak peek and a drive on Ford's test track. Dahlquist noted that scheduled vehicle testing prevented him from setting up timing equipment, but he was allowed to make several banzai runs:

> "Bring the revs up to about a grand on the line. Let the clutch out. Easy on the gas until you're underway and then pin the pedal to the mat. Ahhhhh! All those eight butterflies are flapping open and the sleek Fairlane body is twisting its way to the right, fighting the torque. The tach touched six grand and—wham!—second gear. Then through Third and on to Fourth when you fly over the finish and those 427 inches are still making it all the way through the lights and beyond. No sweat."

Beyond *Hot Rod's* July 1966 article and word of mouth via drag racers, the average Ford guy on the street was not aware of the midyear 1966 Fairlane 427. That changed for 1967 when Ford added both the 410-horse 4V and 425-horse 8V 427s to the Fairlane's option list, including them in the sales brochure as available for "all models except wagons." Other than minor trim updates for the 1967 Fairlane, the 427 package was similar to 1966 with front disc brakes (power for 1967) and handling package.

above: The 1966 Fairlane interior was all business with a bench seat and four-speed shifter. *Mike Mueller*

left: For 1967, the 427 and its optional fiberglass hood were available for all Fairlanes except the station wagon. *Mike Mueller*

The 427 Fairlane...

is also available without numbers.

More and more people who take their driving seriously are turning to the 427 Fairlane. Some of them, like Parnelli Jones and Mario Andretti, go to the trouble of adding personalized touches—such as numbers 18 inches tall!

That's Mr. Jones' 427 Fairlane above, after a Sunday drive that began and ended at Riverside, California recently. Mr. Andretti likes to go Fairlaning near Daytona, Florida.

Before you go to the line next Sunday take a good look at your equipment. If you're serious about your driving, try a real serious car... the 427 cubic inch Fairlane.

This machine produces a no-nonsense 425 horsepower, and a sincere 480 foot-pounds of torque at 3700 revolutions per minute.

If you want to add numbers... go ahead. But, remember, you'll have to give every kid on the block from 8 to 80 a ride!

Mustang • Bronco • Falcon
Fairlane • Ford • Thunderbird • Cortina

For 1967, the Fairlane's flat steel hood was standard equipment; the scooped fiberglass hood shifted to the option list.

With an asterisk explaining "limited production, available for special purchase," and despite the $1,200 price tag—a 50 percent increase over the cost of the car—the 427 found its way into 229 Fairlanes for 1967, including seventy-two high-end 500XLs. Buyers had to sign a Ford disclaimer recognizing the "high performance engine operating characteristics" for use in "supervised competitive events," and the fact that "standard vehicle warranty coverage will not apply."

1967 Comet 427

You can't say that Mercury was hiding the 427 from 1967 Comet customers. Right alongside the 200-cubic-inch six-cylinder and 289 base engines, the sales brochure listed the W-code Cyclone 427 and R-code Cyclone Super 427 big-blocks as options for the Comet 202, Capri, Caliente, and Cyclone. Perhaps the 427's more than $1,200 price tag frightened all but sixty hardcore Mercury performance fans who whipped out their wallets for the hottest Comet ever offered as a street production car. Of those, most were R codes, including six Capris, four Calientes, and nineteen top-of-the-line Cyclones. However, twenty-two hardest of the hardcore knew to order the R-code 427 in the Comet 202 post-sedan, a 9-inch shorter two-door that, at 3,400 pounds and $3,200 base price, offered both weight and cost savings.

Like the 427 Fairlanes, many of the 427-powered Comets were ordered by drag racers. However, they were delivered in full street trim with mufflers, four-speed transmission, and heavy-duty equipment.

opposite: Even though the 427 Fairlane was produced primarily for drag racing, Ford's advertising was quick to connect the dots to the NASCAR Fairlane. "If you want to add numbers, go ahead," said the ad copy. *Donald Farr Collection*

below: The 427 was available for several 1967 Comet models. Most chose the sportier Cyclone. This one has the purposeful "dog-dish" hubcaps in place of the Cyclone's standard chrome covers. *Mike Mueller*

BY THE NUMBERS:

1964 FAIRLANE

ENGINE: 289 High Performance

CARBURETION: Single four-barrel

HORSEPOWER: 271 at 6,000 rpm

TORQUE: 314 at 3,400 rpm

TRANSMISSION: Four-speed

REAR AXLE RATIO: 3.50:1

WEIGHT: 3,100 lbs (curb)

HORSEPOWER TO WEIGHT: 11.43

QUARTER MILE: 16.18 at 87.12 mph (*Car Life,* November 1963)

1964 FALCON SPRINT CONVERTIBLE

ENGINE: 260

CARBURETION: Single two-barrel

HORSEPOWER: 164 at 4,400 rpm

TORQUE: 258 at 2,200 rpm

TRANSMISSION: Four-speed

REAR AXLE RATIO: 3.25

WEIGHT: 3,450 lbs (test)

HORSEPOWER TO WEIGHT: 21.03

QUARTER MILE: 18.00 at 75 mph *(Car Life,* June 1964)

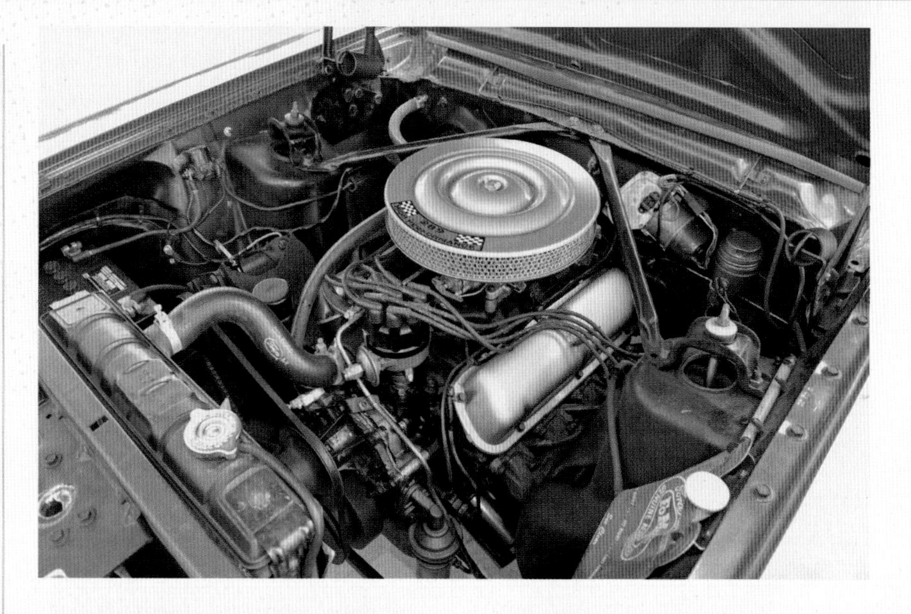

1965 COMET CYCLONE

ENGINE: Super 289

CARBURETION: Single four-barrel

HORSEPOWER: 225 at 4,800 rpm

TORQUE: 305 at 3,200 rpm

TRANSMISSION: Four-speed

REAR AXLE RATIO: 3.50:1

WEIGHT: 3,160 lbs (curb)

HORSEPOWER TO WEIGHT: 15.04

QUARTER MILE: 16.2 at 82 mph (*Motor Trend*, August 1964)

1966 FAIRLANE GTA

ENGINE: 390 Thunderbird Special

CARBURETION: Holley four-barrel

HORSEPOWER: 335 at 4,800 rpm

TORQUE: 427 at 3,200 rpm

TRANSMISSION: Automatic

REAR AXLE RATIO: 3.25:1

WEIGHT: 3,510 lbs (curb)

HORSEPOWER TO WEIGHT: 10.47

QUARTER MILE: 15.2 at 92 mph (*Motor Trend*, October 1965)

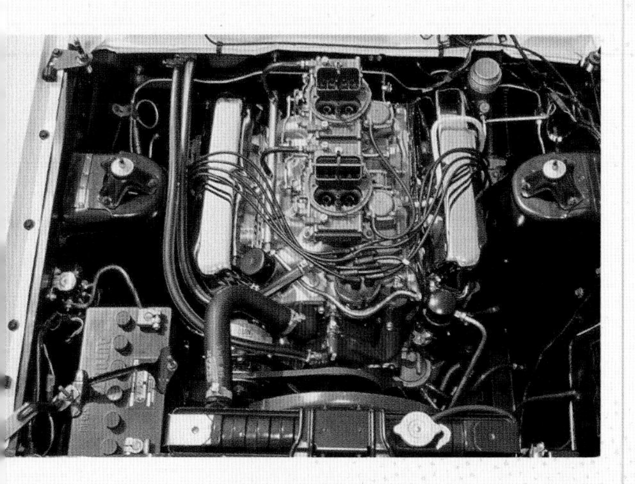

1966 FAIRLANE 427

ENGINE: Cobra 427 8V

CARBURETION: Two Holley four-barrels

HORSEPOWER: 425 at 6,000 rpm

TORQUE: 480 at 3,700 rpm

TRANSMISSION: Four-speed

REAR AXLE RATIO: N/A

WEIGHT: 3,600 lbs

HORSEPOWER TO WEIGHT: 8.47

QUARTER mile: N/A

1966 CYCLONE GT HARDTOP

ENGINE: Cyclone GT 390

CARBURETION: Holley four-barrel

HORSEPOWER: 335 at 4,800 rpm

TORQUE: 427 at 3,200 rpm

TRANSMISSION: Automatic

REAR AXLE RATIO: 3.25

WEIGHT: 3,920 lbs (test)

HORSEPOWER TO WEIGHT: 11.70

QUARTER MILE: 15.2 at 90 mph (*Car Life*, April 1966)

1967 COMET 427

ENGINE: 427 Super Cyclone

CARBURETION: Two Holley four-barrels

HORSEPOWER: 425 at 6,000

TORQUE: 480 at 3,700

TRANSMISSION: Four-speed

REAR AXLE RATIO: N/A

WEIGHT: 3,400 lbs

HORSEPOWER TO WEIGHT: 8.00

QUARTER MILE: N/A

CHAPTER THREE

THE INTERMEDIATES

1968-1973

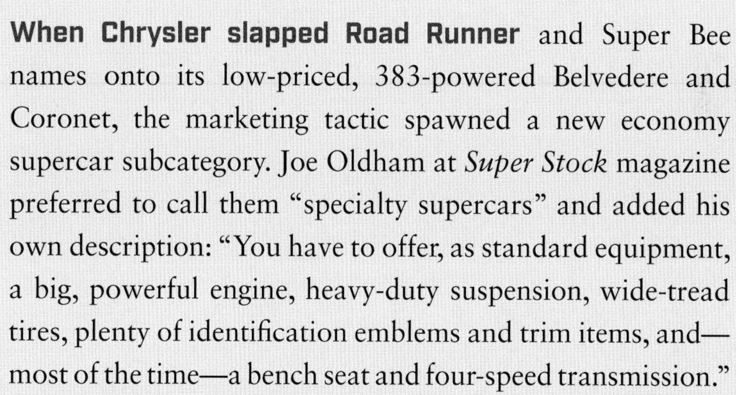

When Chrysler slapped Road Runner and Super Bee names onto its low-priced, 383-powered Belvedere and Coronet, the marketing tactic spawned a new economy supercar subcategory. Joe Oldham at *Super Stock* magazine preferred to call them "specialty supercars" and added his own description: "You have to offer, as standard equipment, a big, powerful engine, heavy-duty suspension, wide-tread tires, plenty of identification emblems and trim items, and—most of the time—a bench seat and four-speed transmission."

Other manufacturers quickly followed suit (even Pontiac) by slicing its GTO into the Judge, a striped and spoilered intermediate named after the "Here Come Da Judge" schtick on the *Laugh-In* comedy show. Ford jumped on the bandwagon for 1969 with the Torino Cobra and Mercury Cyclone CJ. The 428 Cobra Jet supercars gave Ford fans something to brag about.

With the 1969 Cobra, Ford injected a supercar image into its Cobra Jet–powered and Fairlane-based intermediate. *Jerry Heasley*

South Carolina's Ricky Ward was one of those fans, only he was looking up to his older brother, Barry, who bought a 1969 Cyclone CJ, Competition Orange with bench seat, four-speed, and enough CJ decals to send a 383 *Road Runner* running for the hills. "He got it during his senior year of high school," Ricky recalled, "then drove it to Miami when he joined the air force." Later, Ricky bought his brother's Cyclone CJ. "He was known to street race it," Ricky said. "The police would follow me around town thinking it was my brother looking for a race."

The late 1960s were the muscle car heydays for the intermediate Fords and Mercurys, starting with the 428 Cobra Jet at midyear 1968, progressing through the aerodynamic NASCAR-inspired Talladega and Cyclone Spoiler II, and then ending in 1971 after two years of 429 Cobra Jet power. The 351 Cleveland 4V provided the 1972–73 Gran Torino with one last shot of performance before emissions and insurance regulations put an end to the muscle car era.

As for Ricky Ward, he sold the Cyclone CJ for $1,200 in the late 1970s. "Just as well," he said. "Without air conditioning, it was awfully hot driving back from Myrtle Beach."

above: After coveting his big brother's 1969 Cyclone CJ, Ricky Ward ended up buying the Cobra Jet–powered Mercury a few years later. *Donald Farr*

opposite: Because it had no hood scoop and no big Cobra Jet letterings, only simple 428 emblems on the front fenders identified the 428 CJ-powered 1968½ Fairlane or Torino. This one is a Torino GT. *Mike Mueller*

1968½ Fairlane 428 Cobra Jet

On April 15, 1968, Ford's public relations department issued a news release: "The lively image of Ford Division's Mustang and Fairlane takes on an added luster with the release of two 1968½ high-performance packages. Both cars have Ford's recently announced 428-cubic-inch Cobra Jet engine as part of a performance package that places them at the top of the 'supercar' category."

TITUS 3
Firestone

Cobra Jet

More than most Ford dealers, Bob Tasca from Rhode Island's Tasca Ford liked racing and fast cars, realizing—also more than most—that the two went hand-in-hand for increased sales when marketed properly. By summer 1967, Ford's Total Performance commitment had led the blue oval to the top of the racing world in NASCAR, drag racing, Trans-Am, and back-to-back victories on the international stage at LeMans. But the track success was not trickling down to Ford performance enthusiasts. And Bob Tasca wasn't pleased. Tasca performance manager Dean Gregson told *Hot Rod*, "We found the [390] so non-competitive that we began to feel we were cheating the customer. He was paying for what he saw advertised as a fast car but that's not what he was getting. So we did something about it."

That something was to develop a package of existing Ford parts to put customer Fords on par with GM and Mopar muscle cars. To prove his point, Tasca had his dealership technicians install a modified 428 into a 1967 Mustang. With Low Riser 427 heads, a high-lift cam, 427 Fairlane headers, and cold-air feeding a Holley four-barrel on a Police Interceptor aluminum intake, Tasca's Mustang was capable of 13.3-second quarter miles, almost a full second quicker than GTOs and Ram-Air Firebirds. Tasca called it the KR-8—for "King of the Road 1968."

When copies of *Hot Rod* containing Tasca's rant circulated through Ford headquarters, Ford Division Executive Vice President Lee Iacocca demanded a new urgency for promoting performance to the youth market. A resulting November 1967 executive communication laid out a four-year plan—highlighted by big-block Cobra Jet engines—to "dramatize the return to street power." In a closing summary, the memo outlined Ford's new performance priorities: "Put pressure on Product Planning for new image models, increase Engine and Foundry capabilities, clarify and reconsider racing and advertising policies, and do a better job on both street options and hop-up accessories."

"One of our engineers had created a 390 CJ much earlier," explained retired Ford engineer Roger Parlett. "It had the big heads, Police Interceptor intake and carb, and good exhaust. Management saw no future in it because Ford was not into street performance at that time. Thanks to Tasca, the company underwent a mind change."

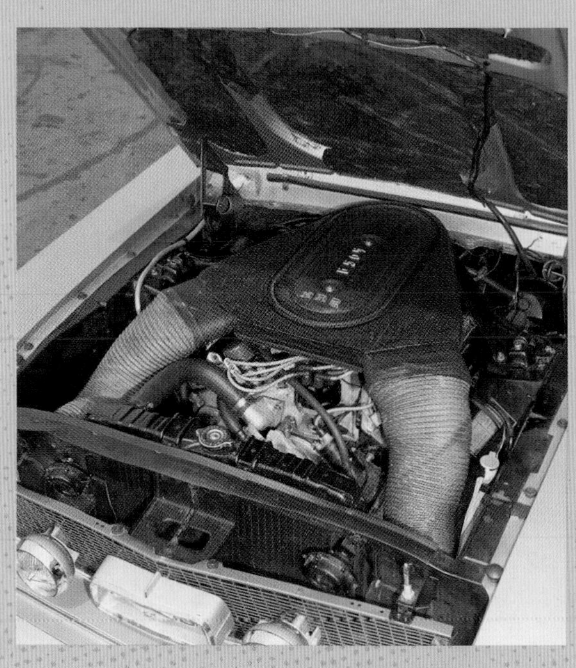

Results came quickly. In February 1968, Ford race teams arrived at the NHRA Winternationals with a new engine in their fastback Mustangs—the 428 Cobra Jet, a name that highlighted Ford's recent acquisition of the Cobra trademark from Shelby. Two months later, in April, the 428 Cobra Jet was announced as an option for the Mustang GT, Cougar, Fairlane/Torino, and Montego/Cyclone.

Using a combination of earlier 390/428 performance engineering and Tasca's KR-8 modifications, Ford based the Cobra Jet on the existing 428 Police Interceptor but increased power with updated 427 Low Riser heads, 390 GT camshaft, cast-iron intake with 735 cfm Holley four-barrel, and low-restriction exhaust manifolds. Ford's conservative 335-horsepower rating was seen as a deception to avoid scrutiny from insurance companies while also gaining an advantage in drag racing. *Hot Rod* noted, "The NHRA bought it lock, stock, and barrel."

At its introduction at midyear 1968, the 428 Cobra Jet was offered as a Ram-Air R code for Mustangs and Cougars and as a Q code without cold-air for the Fairlane and Montego intermediates. For 1969, the new engine was mandatory in the Fairlane Cobra and Cyclone CJ but optional for the Mustang (including the new Mach 1), Cougar, Fairlane, and Montego as either the Q code or R code with Ram-Air, which incorporated a "Shaker" hood scoop for Mustangs.

At midyear 1969, performance enthusiasts received a gift in the form of a new Drag Pack option. Offered for both the Q-code and R-code 428s, the package not only included a 3.91 or 4.30 Traction-Lok differential and engine oil cooler, but also added a more a durable engine reciprocating assembly with cap-screw connecting rods and special balancing for the crankshaft, flywheel, and balancer. These engines became known as Super Cobra Jets, although they were still rated at 335 horsepower. The Detroit Locker was added as a differential choice for 1970.

ENGINE	351-2V	351-4V	390-4V	CJ 428
C.I.D.	351	351	390	428
Adv. H.P. @ RPM	250 @ 4600	290 @ 4800	320 @ 4600	335 @ 5200
Adv. Torque Lb.-Ft. @ RPM	355 @ 2600	385 @ 3200	427 @ 3200	440 @ 3400
Compression Ratio	9.5:1	10.7:1	10.5:1	10.6:1
Carburetor	2-Barrel	4-Barrel	4-Barrel	4-Barrel
Exhaust	Single	Dual	Dual	Dual
Availability:				
Cougar	Std.	Opt.	Opt.	Opt.
Cougar XR-7	Std.	Opt.	Opt.	Opt.
Cyclone CJ	N.A.	N.A.	N.A.	Std.
Cyclone	Opt.	Opt.	Opt.	Opt.
Comet Sports Coupe	Opt.	Opt.	Opt.	Opt.
Montego	Opt.	Opt.	Opt.	Opt.
Montego MX	Opt.	Opt.	Opt.	Opt.
Montego MX Brougham	Opt.	Opt.	Opt.	N.A.

above: An engine oil cooler was a dead giveaway that the 428 was a Super Cobra Jet with heavier-duty internals and 3.91 or 4.30 gearing. *Donald Farr*

inset: In 1969, the 428 Cobra Jet was available in all Mercury Cougars and Montegos except the Brougham, as promoted in this "CJ-428 zip code" ad. *Donald Farr Collection*

opposite left: The earliest 428 Cobra Jet engines had chrome valve covers. Around January 1969, the engine assembly line began installing finned aluminum valve covers. *Tom Shaw*

opposite right: Bob Tasca's KR-8 engine package, built for a 1967 Mustang GT to show what could be done with off-the-shelf performance parts, was eventually tested and examined by Ford during the development of the 428 Cobra Jet. *Archives / TEN: The Enthusiast Network Magazine, LLC*

Cobra Jet continued

1968½ Cyclone 428 Cobra Jet

Perhaps *Motor Trend*'s Eric Dahlquist had watched too many *Batman* episodes when he described the acceleration characteristics of the 1968½ Cyclone GT powered by the new 428 Cobra Jet:

> "You mash the accelerator pedal and the Merc leaps away from the mark like you had backed into a big coil spring. Wham! Chirp! That beautifully positive transmission socks you right between the shoulder blades and the wide F70x14 Goodyear Polyglas tires let out a yell. Wham! Chirp! There it goes again and the speedometer looks like it's spring loaded as it sweeps past 100 mph."

Matching the mid-1968 Fairlane, Mercury offered the 428 Cobra Jet in its intermediates, which were restyled for 1968 and renamed Montego, with Comet and Cyclone sub-models. The new 335-horsepower big-block replaced the 427, which had been listed in the sales brochure but never materialized in a production midsize Mercury.

With the Comet name downgraded to a base two-door Montego, the Cyclone entered 1968 as the performance-oriented intermediate, available as hardtop or fastback. The base engine was the 302 two-barrel with powertrain options up to the 390 four-barrel, while adding the GT package provided bucket seats, upper and lower stripes, and a special performance handling package with 5½-inch wheels and turbine wheel covers.

"Mind blowing, brute acceleration is the 428 Cyclone's long suit," said *Motor Trend* in its road test. "Right off the street, it ran 14.39 at 98 mph through Orange County Raceway clocks. A few more passes and it was pinching 14.12s at 99 and change. Driving the 428 Cyclone GT will warm the hearts of the combined drag- and road-race fraternity."

While the 1968 Cyclone GT was the ideal platform for the 428 CJ, the new engine was also available in the Montego hardtop and Comet Sports Coupe. Unlike the 1968½ Cobra Jet Mustang and Cougar that came standard with Ram-Air, the intermediate Mercurys got the Q code without cold air. With the CJ under the hood, the midsize Mercurys were also equipped with heavy-duty suspension and 9-inch rear end. Only 186 1968½ Cyclones were produced with the 428 Cobra Jet during their short production cycle.

After testing the 1968½ Cyclone with 428 Cobra Jet power, writer Eric Dahlquist exclaimed, "Here come the jets like a bat out of hell!" *Archives / TEN: The Enthusiast Network Magazine, LLC*

With hood scoop, blackout grille, and snake emblems, the 1969 Fairlane-based Cobra finally gave Ford's intermediates a supercar image. *Jerry Heasley*

The 1968 Fairlane that appeared in Ford dealer showrooms was considerably different from the 1967s. Growing once again in size and weight, the Fairlane gained a more formal appearance for the hardtop and a sleeker look for the new fastback with a roofline that stretched from the top of the windshield to the rear panel. There was also a new subseries of Fairlane called the Torino, essentially the same car but with a more upscale image thanks to the Italian name (for the city of Turin) and additional trim. One magazine described the Torino as a "gussied-up" Fairlane. The Torino was offered as a GT, but it was mainly an appearance package with stripes and styled steel wheels. The Torino GT was a sales success with more than one hundred thousand sold, but most were powered by the standard two-barrel 289 or 302.

The Fairlane entered 1968 with the previous year's 390 Thunderbird Special, uprated with 5 more horsepower to 325, as an option for all models. A 390-horse 427 was also listed, but it was quickly dropped and none were built, perhaps because Ford Engine Engineering was working on a new performance big-block, one based on the regular production 428 and therefore less expensive to produce.

Announced midyear at $306 over the cost of the base V-8, the 335-horsepower 428 Cobra Jet was available for all 1968 Fairlanes except the station wagon. It was offered with an upgraded Cruise-O-Matic C6 transmission and 3.50 rear axle, with locking axle optional for 3.50 or 3.91 gearing. While 1968½ Cobra Jet Mustangs and Cougars came with a functional hood scoop and Goodyear's new Polyglas tires, the CJ-powered 1968 Fairlanes kept their stock flat hoods and Wide Oval rubber. The competition suspension was a mandatory option for CJ cars.

Car & Driver got some seat time in a 1968 CJ Torino GT during a comparison with David Pearson's NASCAR Torino. After noting that the press car was equipped with a

The 428 Cobra Jet was available optionally in all 1969 Fairlanes, including the Torino GT. *Eric English*

preview of 1969's functional hood scoop, the magazine reported a 14.2-second quarter mile at 98.9 miles per hour, more than a full second improvement over *Car Life's* earlier test with a 390 Torino. "The 1-2 shift broke the Wide-Ovals loose for at least a length," *C&D* reported. "Ford lovers have a reason to rejoice."

According to Marti Auto Works's Ford production database, 840 owners rejoiced by ordering the 428 Cobra Jet for their 1968½ Fairlane.

1969 Cobra

Ford delivered its first volley into the budget supercar ranks with the 1969 Cobra, a no-frills Fairlane 500 (by body code, although neither the Fairlane or Torino name appeared on the car) that provided a muscle car image to back up its standard 428 Cobra Jet powerplant. Offered as a two-door fastback or hardtop, early Cobras even parodied the *Road Runner*'s cartoon character image; instead of a roadrunner bird in a cloud of dust, the Cobra's decal was a snake on wheels. Except Ford didn't have to pay a royalty to Warner Brothers.

In 1967, Ford had acquired the Cobra trademark from Carroll Shelby, reportedly for one dollar. For 1969, Ford doubled-down on the name recognition by making the 428 Cobra Jet the standard engine for the new Cobra midsize supercar.

In its intended low-buck form, the Cobra retailed for $3,139 and came standard with the Q-code non–Ram-Air 428 Cobra Jet engine with chrome (early) or aluminum (beginning early 1969) valve covers, four-speed transmission, 9-inch rear end with 3.25:1 gearing, 6-inch styled steel wheels with F70x14 tires, heavy-duty battery, cooling package, dual exhaust, and competition suspension, with four-speed cars equipped with staggered rear shocks. Externally, the Cobras were identified by their blackout grille, hood pins, and identification, with the aforementioned cartoon decal quickly replaced by a simple coiled snake emblem over Cobra lettering.

The base Cobra was a great starting point. For $133, Cobra buyers could upgrade to the R-code 428 Cobra Jet with Ram-Air, which fed cooler, denser outside air into the engine via the hood scoop, although the power rating remained the same at 335. For harder launches, optional gearing included 3.50 and 3.91, with Traction-Lok at extra cost. Bucket seats and in-dash tachometer were also available. Even the beefed-up Select-Shift automatic was a great choice for the 428 Cobra Jet; it easily barked the rear tires on the second and third upshifts at wide-open throttle. When optioned with the console, the automatic's shifter moved from the steering column to the floor.

While the 428 Cobra Jet was the only available engine for the Cobra, both the Q-code and R-code 428s were offered in other Fairlanes and Torinos, even the pickup-like Ranchero. R-code cars added the functional hood scoop with chrome "428 Cobra Jet" lettering. Q-code cars, including Cobras, were delivered with a flat hood unless the scoop was ordered separately as nonfunctional dress-up.

By spring 1969, Ford had added a Drag Pack option for 428 Cobra Jet vehicles. To alleviate overheating issues caused by high engine speeds with 3.91 and 4.30 gearing, the package added an engine oil cooler mounted in front of the radiator, along with Traction-Lok differential. Available for both the Q-code and R-code engines, the Drag Pack also upgraded the engine to Super Cobra Jet status with cap-screw connecting rods and modified crankshaft.

Interestingly, Ford referred to the 1969 Cobra as the *Boss Snake* in this two-page ad spread. The Boss name would show up later in the year on a pair of Mustangs. *Donald Farr Collection*

left: The early cartoonish snake decal was replaced by a coiled snake emblem. *Jerry Heasley*

right: The "Cobra Jet 428" emblem on the scoop left no doubt about what was under the hood. The scoop was supplied only with the R code 428 Cobra Jet for functional Ram-Air. *Jerry Heasley*

opposite top: The 1969 Cyclone CJ was loaded with image—decals front and rear, blackout trim, and dual exhaust. This one has the optional styled steel wheels in place of the standard wheel covers. *Jerry Heasley*

opposite bottom: Interior options for the Cyclone CJ included bucket seats, console with floor shift for automatic, and tachometer instrument cluster. *Jerry Heasley*

Naturally, the car magazines leaped at the chance to test the new Cobra as competition heated up between the budget supercars, with some pitting the various models against each other. Writing for *Super Stock*, Joe Oldham demonstrated his way with words when he reported on a Cobra SportsRoof with automatic and 3.50 Traction-Lok:

> "Say you're cruising along at 60 and you want to pass. First, check your left racing mirror. Pull out into the left lane and check ahead. All clear. Now, floor it! Bang! Zoom! WaaaahhhHHH! So many things happen at once that it's hard to keep track. First of all, the automatic transmission downshifts into second gear. And at 60 mph, the 428 cubic inches of Cobra Jet is producing enough torque to push you back into the seat so hard that you feel like someone hit you in the chest with a medicine ball. Top end power? Unreal. Enough to push you to at least 125 mph with the fastback roof and 3.50 gears. At least that's where we quit."

Oldham recorded a 14.8-second quarter mile, while *Car Life* clocked a 14.9 and *Motor Trend* nailed a 14.5. However, *Car & Driver* topped them all with a 14.04-second ET at 100.61 miles per hour. It even made a pass with the Ram-Air blocked and noted a 0.2-second slower time. "The Cobra Jet has a definite appetite for fresh air," a review said.

For a specialty supercar, the 1969 Cobra sold well at 14,885: 11,099 as fastbacks and 3,786 as hardtops.

1969 Cyclone CJ

With the 428 Cobra Jet added to its stable of engines, Mercury stepped out of its luxury comfort zone with a new performance campaign for 1969. Ford's corporate cousin even created its own youthful "streep" language—"pertaining to street/strip," explained the "Streep Scene" brochure, with colorful psychedelic graphics. "Streep sleeper," it said, stood for "Doesn't look like it can, but it will."

As with Ford's Fairlanes, the 428 Cobra Jet was optional in all 1969 Montego-based Mercurys except the luxury MX Brougham. Similar to the Fairlane Cobra, the Cyclone

MERCURY
331YAC

opposite top: Designed for the NASCAR speedways, the 1969 Talladega sliced through the air with its flush-fit nose. The front bumper was actually a modified rear bumper. *Mike Mueller*

opposite bottom: Starting out as base 1969 Cobras, the Talladegas were equipped with black bench-seat interiors with column shift for the mandatory automatic transmission. *Mike Mueller*

CJ showcased the new 428 as a specialty supercar, a fastback only and easily identified with blackout grille, twin black or silver hood stripes, and can't-miss-'em decals, "CJ 428" on the front fenders and "Cyclone CJ" on the rear quarters. As a budget muscle car, the Cyclone CJ came with a four-speed, a 9-inch rear axle with 3.50:1 gears, Competition Suspension (with staggered rear shocks for manual trans), low-restriction dual exhaust, plain wheel covers, and wide-tread white sidewall tires. The $3,000 base price was attractive, but by the time buyers checked off optional Ram-Air (which added a black hood scoop, hood pins, and Goodyear Polyglas tires), automatic transmission, bucket seats, tachometer, 3.91 or 4.30 gears, Traction-Lok differential, styled wheels, power front disc brakes, and a few other goodies, the sticker price could easily top $4,300. Cheap or loaded, Mercury sold only 2,175, according to Marti Auto Works Ford production database.

"The Cyclone CJ can best be described as a gentleman's muscle car," said *Car & Driver* during a test with four other budget supercars. Equipped with the Ram-Air engine and 3.91 Traction-Lok, *C&D*'s 3,860-pound test car tripped the quarter mile lights at 13.94 seconds at nearly 101 miles per hour, topping a 3.50-geared Cobra by a full second.

Other magazine drag tests resulted in similar quarter-mile performances—13.99 at 103.92 miles per hour by *Super Stock* with a 4.30-geared CJ and 14.05 at 101 miles per hour by *CARS* magazine. "The Mercury Cyclone 428 CJ has the unique distinction of being one of the most 'in' cars on the supercar scene," said *CARS*. "It looks like it's breaking the speed limit when it's idling." *Super Stock* added, "The thing we liked most about the Cyclone? Easily the 428 Cobra Jet engine, which will go down in history as one of the strongest wedges ever engineered."

1969 Talladega

By 1968, NASCAR had become an important marketing tool for American auto manufacturers. Winning on Sunday meant selling more cars on Monday. During the mid-1960s, it was all about engines—Chrysler 426 Hemi versus 427 Fords and canted-valve 427 Chevys. Aerodynamics entered the picture in 1968 when the Fairlane fastback's sleek shape proved advantageous on the superspeedways. In 1969, Chrysler fired back with a slippery Charger 500, building just enough production models with flush rear window glass and blunt front end to satisfy NASCAR's 500-car homologation rule. Ford countered at mid-1969 with the Talladega, a flush-nose Cobra fastback named for the town where Bill France was building the 2.66-mile Alabama International Motor Speedway. The track wouldn't open until fall 1969, but it was already predicted as the world's fastest closed-course racetrack.

While Ford's contracted Kar Kraft assembly line was churning out Boss 429 Mustangs to legalize a new hemi-head race engine, Ford's Atlanta Assembly Plant was converting 1969 Cobras into Talladegas. The process was not simple or inexpensive; the original front end was cut off and a nearly 6-inch longer extension grafted on. The factory grille was moved forward and sealed with a rubber gasket, creating a flush fit for improved front-end aerodynamics. A fabricated panel filled in the gap between the fenders in front of the hood. The front bumper was actually a rear bumper that had been cut, reshaped, and pieced back together to serve as both protection and air spoiler. Less obvious were the rerolled rocker panels, a trick that raised the rocker height by 1 inch to allow the race cars to sit lower on the track.

FORD
T
Talladega
Ford
1969 NASCAR GRAND NATIONAL CHAMPION

Ford President Bunkie Knudsen was a big NASCAR fan, so Kar Kraft built a special Talladega for him. Production Talladegas were available in only three colors, but Knudsen's was painted in a special shade of yellow. *Jerry Heasley*

The 1969 Cyclone Spoiler came in two versions—Cale Yarborough with red trim (pictured) and Dan Gurney with blue trim. *Dale Amy*

To keep things simple and low cost, all Talladegas started life as base Cobra fastbacks with black bench-seat interior, automatic transmission (with steering column shifter), and 3.25:1 non-locking rear axle. Like all 428 Cobras, the Talladegas came with the Competition Suspension, but they were the only CJ-powered cars to use the staggered rear shock arrangement with automatic transmission. An oil cooler was also standard equipment. Colors were restricted to three—Wimbledon White, Presidential Blue, and Royal Maroon—with black hoods and rear panels. Identification was limited to cast "T" plates over the exterior door handles and a "T" emblem gas cap. Because it was impossible to reach the hood latch through the closed grille, a cable-operated remote hood release handle was mounted under the instrument panel.

Although required to build only 500, Ford produced 750 Talladegas, often delivering them to Ford dealers who had no idea how to promote them. Ford never promoted or advertised the Talladegas other than a simple four-page sales brochure.

In NASCAR, the aerodynamic Talladega convinced Plymouth stalwart Richard Petty to switch to Ford for the 1969 season. His Petty Blue Talladega won ten races, one fewer than David Pearson in a Holman-Moody Talladega. Pearson won the 1969 championship, with Petty second.

1969 Cyclone Spoiler/Spoiler II

To homologate engines and aerodynamic aids for NASCAR, Ford was producing Boss 429-powered Mustangs and snub-nosed Talladegas. Mercury took a slightly different approach by promoting its NASCAR activities with a pair of special Cyclones, one named for East Coast driver Cale Yarborough and the other for West Coast racer Dan Gurney. Announced in January 1969, they were available as the Cyclone Spoiler with the standard front end or as the Spoiler II with an extended flush-fit nose.

Based on the regular-production Cyclone, the Spoilers featured exactly that—a pedestal-mounted spoiler on the rear trunk lid. All Cyclone Spoilers were white with two-tone blue or red roofs and upper side stripes to match the NASCAR Mercurys campaigned by Yarborough (red) or Gurney (blue). Gold with black decals, either "Cale Yarborough Special" or "Dan Gurney Special," were placed on the front fenders. A package deal, the Cyclone Spoilers were also equipped with blackout grille, hood scoop, black hood stripes, hood pins, argent styled steel wheels, and F70x14 raised white-letter Goodyear Polyglas tires. The standard engine was the 290-horsepower 351 four-barrel with the 390 and both 428 Cobra Jets as options. All other Cyclone options were available, from bucket seats to air conditioning.

The second version of the 1969 Cyclone Spoiler served the same purpose as Ford's Talladega to homologate a more aerodynamic front end for Mercury's NASCAR teams. Although the Spoiler II nose appeared the same as the Talladega, it was 4 inches longer with a 35-degree slope instead of the Talladega's 30-degree angle. Like the Talladega, the grille was moved forward and sealed with a rubber gasket (forcing the use of an under-dash cable-operated hood release), the front bumper was a modified rear bumper, and the rocker panels were rerolled for a height advantage on the track. The production Spoiler IIs included the dealer-installed rear spoiler but did not come with the hood scoop and hood pins like the standard Cyclone Spoiler. Also like the Talladegas, the Spoiler IIs were optioned identically, unfortunately with the 290-horsepower 351 four-barrel V-8. The 428 Cobra Jet was not available.

Mercury built 969 Cyclone Spoilers, 617 in Cale Yarborough white/red colors and 352 as white/blue Dan Gurneys. Only 98 were ordered with the 428 Cobra Jet. Spoiler II production barely exceeded NASCAR's 500-car requirement, with 503 built, 285 as Cale Yarborough cars and 218 as Dan Gurneys.

Of course, the 1969 Cyclone Spoiler was equipped with a spoiler, pedestal-mounted on the trunk lid. *David Newhardt*

1970–71 Cyclone Spoiler 429 Cobra Jet

As before, the 1970 Cyclone and Montego shared its unibody, mechanicals, and powertrains with Ford's Torino. But the Mercury intermediate was a totally different animal, starting up front with a protruding snout that made the car look even longer than the 5 inches added over the 1969 model. *Hot Rod* described it as "less of a Ford and more of a Mercury."

As the Mercury muscle model, the $3,200 Cyclone Spoiler's standard engine was the 370-horsepower 429 Cobra Jet Ram-Air, not the 429 Thunder Jet, like the Cobra. Like the Ford version, the 429 CJ was converted into a 375-horsepower Super Cobra Jet when ordered with the 3.91 (Drag Pak, in Mercury's spelling) or 4.30 (Super Drag Pak) gearing.

Hot Rod flogged a 429-powered 1970 Cyclone Spoiler to a 14.23-second quarter mile. About the new front end, it said, "The extended nose could be a vulnerable part in a parking lot." *Archives / TEN: The Enthusiast Network Magazine, LLC*

True to its name, the 1970 Cyclone Spoiler had spoilers, two of them: one a large, black air dam underneath the front bumper and the other an adjustable, pedestal-mount wing on the trunk lid. With Ram-Air standard, the hood scoop was functional via a vacuum-operated flapper valve assembly on the air cleaner. The blackout grille incorporated a "gun sight" snout with hideaway headlights optional. Competition Suspension with G70x14 tires, a 3.50 differential, 8,000 rpm tachometer, and Hurst-shifted four-speed (with staggered rear shocks) were standard. The Cruise-O-Matic was optional. Identification included Cyclone Spoiler lettering on the front fenders trailed by a long tape stripe and block Spoiler decal on the trunk lid.

For 1971, like Ford's Cobra, the Spoiler's standard engine became the 285-horsepower 351 Cleveland with the 429 Cobra Jet as an option, although the Drag Pak disappeared. Externally, the big difference was the side stripe: for 1971, the Cyclone Spoiler lettering was placed forward of the front wheel openings with a much bolder stripe that ran along the upper body line. The rear spoiler, painted body color in 1970, was black for 1971.

During 1970 and 1971, just 1,827 Cyclone Spoilers came with the 429 Cobra Jet, 1,631 for 1970 and 196 for 1971.

above: The Cyclone Spoiler also featured a rear deck spoiler, painted black for 1971. Mounted on a pair of trunk lid pedestals, the wing angle was adjustable by loosening a pair of Allen-head screws, as designed by Ford stylist Larry Shinoda for the 1969 Mustang (and noted on the vanity license plate). *Jerry Heasley*

left: The 429 Cobra Jet was optional in the 1971 Cyclone Spoiler, still rated at 370 in the last year of true Ford performance engines. *Jerry Heasley*

1970–71 Torino Cobra

> *Car & Driver*:
>
> "A quick look to your right and you see that the driver in the other lane is nervous . . . inching forward, hoping to get a few inches advantage. The light goes amber, you bring up the revs . . . not too high, about 1,500 is enough to launch you cleanly with those big, tricky Polyglas tires. Your hand tightens on the Hurst tee-bar shifter. Green. You're on it instantly, a bit too much and the rear end starts to come around, but you catch it and see out of the corner of your eye that you've already got a fender on the car next to you. Pow! At 6,000 rpm, you stab at the clutch and haul back on the shifter. Pow! You almost ram the tee-bar through the dashboard and you can't even see the other car anymore."

As was customary for the era, the Fairlane saw styling and mechanical updates every two years. But never was change so radical as the Fairlane's transition into the 1970s. In fact, the Fairlane name—a staple in the Ford lineup since the 1950s—was downgraded to the "modestly priced" Fairlane 500 as a subset to the Torino, the new preferred identity for Ford's 1970 intermediate.

Except the all-new Torino was more of a full-size than an intermediate. Gaining an inch in wheelbase (up to 117) and 5 inches in overall length (206.2), the 1970 Torino was nearly the same size as the early 1960s' Galaxies. Explained *Car & Driver*: "What would have passed for a full-sized Ford hardtop in 1964 was a 3,700-pound car with a 119-inch wheelbase and 209 inches overall length. Ford's [Torino Cobra] has a shipping weight of over 3,900 pounds."

Offsetting the weight gain was a new big-block engine. While the Mustang clung to the 428 for one more year, the 1970 Cobra moved into the future with the 429 from Ford's new 385 series engine family, which used cylinder heads with canted valves and large round ports. In base form, the Cobra's N-code 429 Thunder Jet developed 360 horsepower. But to really take advantage of the Cobra's supercar image, buyers could choose one of two optional 429s—the C-code Cobra Jet or the J-code Cobra Jet Ram-Air, both rated at 370 horsepower. With Ram-Air, the air cleaner was topped by a functional through-the-hood scoop, similar to the Shaker found on the Mustang but reshaped for the Torino. Ordering the Drag Pack's 3.91 or 4.30 gearing automatically updated the 429 Cobra Jet to the 375-horsepower Super Cobra Jet with solid-lifter cam, Holley four-barrel, oil cooler, and beefier internals.

As the Torino's supercar offering, the 1970 Cobra SportsRoof came standard with Hurst-shifted four-speed and Competition Suspension with 7-inch wheels and hubcaps and F70x14 raised white-letter tires, plus staggered rear shocks for four-speed cars. Externally, the Cobra flexed its muscles with a flat-black hood, blackout grille, new twist-type external hood latches, and Cobra decals on the fenders and rear panel. Ford sold 7,675 Torino Cobras for 1970.

All three 429s were available in other 1970 Torinos, including the Torino GT and even the Falcon, reintroduced at mid-1970 as a base Torino.

Motor Trend pitted a 1970 Torino Cobra against a 454 Chevelle and 440 Road Runner, with the Ford coming up on the short end of the quarter-mile stick with a 14.5 clocking versus 13.8 for the 450-horse Chevy and 14.4 for the six-barrel Mopar. "The Torino Cobra was the biggest of our trio," the editors explained. "The engine was the mildest of the three tested, yet the car had the quickest 0–30, 0–45, and 0–75

"Shaped by the wind," per the marketing copy, the 1970 Torino Cobra was based on the long, low, and sleek new SportsRoof body. *Eric English*

times, and matched the Chevelle's six seconds to 60 miles per hour. It also ran the highest quarter-mile speed, 100.2, a fact which is even stranger in view of its weight."

In a precursor of things to come, the 1971 Cobra's standard engine was downgraded to the 285-horsepower 351 Cleveland, and the 360-horse 429 Thunder Jet disappeared, although both 429 Cobra Jets—called CJ and CJ-R with Shaker hood scoop—were offered optionally for 370 horsepower, along with the 375-horsepower Super Cobra Jet when ordered with the Drag Pack. CJ-powered Cobra production for 1971 dropped to 968.

1972 Gran Torino Sport

The differences between the 1971 and 1972 Torinos were dramatic, both externally and under the hood.

In 1971, the Torino's 351 Cleveland 4V engine was rated at 285 horsepower. Due to lower compression ratios and a new "net" method of measuring advertised engine horsepower (installed in the car with accessories, emissions equipment, and exhaust system as opposed to the previous "gross" measurement), the 1972 rating dropped to 248 for the Q-code 351 4V, which was also called the Cobra Jet. (Due to differences in emissions and exhaust systems, the same engine was rated at 266 horsepower in the Mustang).

The 1972 Torino was also an altogether different car than 1971. Abandoning the muscular look, Ford took the Torino upmarket by completely restyling the body with a more pronounced "Coke-bottle" profile and prominent oval grille. Underneath, the 1972 Torino was reworked with body-on-frame construction and coil-spring suspension, with two-door models getting their own shorter 114-inch wheelbase. The 429 made a final appearance, but it was the run-of-the-mill variety with 205 horsepower, not the Cobra Jet from the year before, and more suited for towing trailers.

For performance fans, the Gran Torino Sport was the hot ticket and the last vestige

Totally new styling propelled the 1972 Gran Torino into the 1970s. An available 351 4V engine with Ram-Air was the last vestige of muscle car performance. *Jerry Heasley*

429 Cobra Jet/ Super Cobra Jet

Adding more confusion to its 427 and 428 displacements, Ford introduced a 429-cubic-inch big-block in 1968, initially available for the Thunderbird before joining the powertrain list for all full-size Fords in 1969. Marketed as the Thunder Jet, the 429 was part of Ford's new 385 series, so-named for the largest displacement's (460 cubic-inch) 3.85-inch stroke in a new thin-wall, skirtless block. The cylinder heads were a canted-valve design that allowed larger ports and valves. A beefed version of the 429's short-block became the foundation underneath the hemi heads for the 1969–70 Boss 429 Mustangs.

While the N-code 429 Thunder Jet's 360 horsepower was impressive thanks to a 11.0:1 compression ratio and a 600 cfm four-barrel carburetor, the two Cobra Jet versions boasted 370 horsepower with 11.3:1 compression, larger intake and exhaust valves, hotter camshaft, header-type exhaust manifolds, and a 760 cfm Rochester Quadrajet four-barrel on a high-rise intake. The 429 CJ was available as a C code or as the J-code Ram-Air with functional cold-air induction, which used a Shaker scoop for the Torino or NASA-style twin hood ducts when the 429 became available in the Mustang for 1971.

Like the earlier 428 Cobra Jet, both 429 CJs were upgraded to Super Cobra Jets when ordered with the Drag Pack, which provided 3.91 Traction-Lok or 4.30 Detroit Locker gears while also enhancing the engine with forged aluminum pistons, four-bolt main bearing block, solid-lifter cam, and 780 cfm Holley four-barrel carburetor. Strangely, even with the improved breathing capabilities, Ford rated the 429 Super Cobra Jet at 375 horsepower, only five horsepower more than the standard Cobra Jets.

above: The 429 Cobra Jet was optional for the 1971 Cyclone Spoiler, which lived up to its name with a huge air dam at the front and a wing on the trunk deck. *Jerry Heasley*

left: Based on Ford's 385-series big-block, the 429 Cobra Jet was rated at 370 horsepower in both Ram-Air and non-Ram-Air form. This one has the Drag Pack option, making it a Super Cobra Jet with 375 horsepower. *Eric English*

Swoopy fastback styling was popular for the 1972 Gran Torino, especially in SportsRoof form with optional Magnum 500 wheels and Laser stripes. *Jerry Heasley*

of the Torino's muscle car past. The SportsRoof provided swoopy fastback styling, but it was the option list that transformed the Gran Torino from grocery-getter to performance car. Although the compression ratio and horsepower rating dropped, the available 351 Cleveland 4V was basically the same as before with large intake ports and 750 Motorcraft four-barrel. It was the only Torino engine available with a four-speed and dual exhaust, which had its own sound. All the good stuff came with the optional Rallye Equipment Group, which packaged the 351 4V engine, four-speed with Hurst shifter, Competition Suspension, 14x6-inch wheels with white-letter tires, and eight-pod gauge package with tachometer. Adding functional Ram-Air hood and Traction-Lok differential helped the Gran Torino Sport achieve mid–15-second ETs, while Magnum 500 wheels and Laser side stripes provided an image to match.

Car & Driver described the 1972 Gran Torino Sport as "quiet as a Jaguar, smooth as a Continental." They weren't impressed with the emasculated power: "The Torino's brawny sheetmetal contours and potent-looking rubber can have you believing the car is mighty enough for Daytona. But the power under the hood says that acceleration has tumbled down Ford's list of priorities."

CARS magazine clocked a 15.40-second quarter mile but refused to put the Gran Torino Sport into the supercar category. "Properly optioned," it said, "the Torino is a very neat highway hauler, offering respectable handling, quiet ride, and acceptable performance."

The Q-code 351-powered Gran Torino was a popular car for 1972, with 15,688 sold.

For 1973, the Gran Torino adopted new federally mandated 5-miles-per-hour bumpers, which gave the front end more of a blunt appearance. The 351 4V was further detuned with smaller valves. Ford made no reference to the horsepower rating in its sales and advertising material.

BY THE NUMBERS:

1969 COBRA SPORTSROOF

ENGINE: 428 Cobra Jet Ram-Air

CARBURETION: 735 cfm Holley four-barrel

HORSEPOWER: 335 at 5,200 rpm

TORQUE: 440 at 3,400 rpm

TRANSMISSION: Automatic

REAR AXLE RATIO: 3.50:1 Limited-slip

WEIGHT: 3,890 lbs (curb)

HORSEPOWER TO WEIGHT: 11.61

QUARTER MILE: 14.04 at 100.61 mph (*Car & Driver*, January 1969)

1968½ TORINO GT FASTBACK

ENGINE: 428 Cobra Jet

CARBURETION: 735 cfm Holley four-barrel

HORSEPOWER: 335 at 5,400 rpm

TORQUE: 440 at 3,400 rpm

TRANSMISSION: Automatic

REAR AXLE RATIO: 3.91

WEIGHT: 3,362 lbs (curb)

HORSEPOWER TO WEIGHT: 10.03

QUARTER MILE: 14.2 at 98.9 mph (*Car & Driver*, June 1968)

1968½ CYCLONE GT

ENGINE: 428 Cobra Jet

CARBURETION: 735 cfm Holley four-barrel

HORSEPOWER: 335 at 5,600 rpm

TORQUE: 445 at 3,400 rpm

TRANSMISSION: Automatic

REAR AXLE RATIO: 4.11 Traction-Lok

WEIGHT: 3,880 lbs (curb)

HORSEPOWER TO WEIGHT: 11.58

QUARTER MILE: 13.86 at 101.69 mph (*Motor Trend*, August 1968)

BY THE NUMBERS:

1969 CYCLONE CJ

ENGINE: 428 Cobra Jet Ram-Air

CARBURETION: 735 cfm Holley four-barrel

HORSEPOWER: 335 at 5,200 rpm

TORQUE: 440 at 3,400 rpm

TRANSMISSION: Automatic

REAR AXLE RATIO: 3.91:1 Limited-slip

WEIGHT: 3,860 lbs (curb)

HORSEPOWER TO WEIGHT: 11.52

QUARTER MILE: 13.94 at 100.89 mph (*Car & Driver*, January 1969)

1969 CYCLONE SPOILER

ENGINE: 428 Cobra Jet

CARBURETION: 735 cfm Holley four-barrel

HORSEPOWER: 335 at 5,200 rpm

TORQUE: 440 at 3,400 rpm

TRANSMISSION: Automatic

REAR AXLE RATIO: 3.25:1

WEIGHT: 3,860

HORSEPOWER TO WEIGHT: 11.52

QUARTER MILE: N/A

1969 TALLADEGA

ENGINE: 428 Cobra Jet

CARBURETION: 735 cfm Holley four-barrel

HORSEPOWER: 335 at 5,200 rpm

TORQUE: 440 at 3,400 rpm

TRANSMISSION: Automatic

REAR AXLE RATIO: 3.25:1 non-locking

WEIGHT: 3,775 lbs (curb)

HORSEPOWER TO WEIGHT: 11.26

QUARTER MILE: N/A

1970 TORINO COBRA

ENGINE: 429 Super Cobra Jet Ram-Air

CARBURETION: 780 cfm Holley four-barrel

HORSEPOWER: 375 at 5,200 rpm

TORQUE: 450 at 3,400 rpm

TRANSMISSION: Four-speed

REAR AXLE RATIO: 3.91:1 Traction-Lok

WEIGHT: 3,586 lbs

HORSEPOWER TO WEIGHT: 9.69

QUARTER MILE: 13.99 at 101 mph (*Motor Trend*, February 1970)

1970 CYCLONE SPOILER

ENGINE: 429 Cobra Jet Ram-Air

CARBURETION: 760 cfm Quadrajet four-barrel

HORSEPOWER: 370 at 5,400 rpm

TORQUE: 450 at 3,400 rpm

TRANSMISSION: Automatic

REAR AXLE RATIO: 3.50 Traction-Lok

WEIGHT: 3,980 lbs (test)

HORSEPOWER TO WEIGHT: 10.75

QUARTER MILE: 14.23 at 101.12 mph (*Hot Rod*, April 1970)

1972 GRAN TORINO SPORT

ENGINE: 351 Cleveland 4V

CARBURETION: Motorcraft four-barrel

HORSEPOWER: 248 at 5,400 rpm

TORQUE: 299 at 3,600 rpm

TRANSMISSION: Automatic

REAR AXLE RATIO: 3.50:1 Limited-Slip

WEIGHT: 3,800 lbs (curb)

HORSEPOWER TO WEIGHT: 15.3

QUARTER MILE: 15.40 at 86 mph (*CARS*, December 1972)

1965

CHAPTER FOUR

PONY CARS I

1965-1968

Herb Gordon describes his father as a "hot-rodder and a racer." In 1965, Bob Gordon ordered a Mustang fastback with the 289 High Performance and four-speed. But the factory 271 horsepower wasn't enough for the West Virginia thrill seeker, so he ordered Ford's over-the-counter Cobra II 4V Induction kit with its pair of 500 cfm Carter AFB four-barrel carbs. Dipping his right toe into eight barrels of fun proved too tempting; Bob's driving habits soon attracted the attention of local law enforcement. As a car dealer, Bob couldn't risk losing his driver's license, so he parked his fastback after driving it less than three thousand miles.

Before the Mustang, buyers shopping for a performance Ford were limited to the big Galaxie, the boxy Fairlane intermediate, or economy-based Falcon Sprint. Although the Mustang was based on the Falcon, sharing its unibody, suspension, steering, and instrument cluster, the long-hood, short–rear deck styling provided the Mustang with a sports

In 1965, Bob Gordon purchased a new Mustang fastback with the 289 High Performance. He passed it on to his son Herb, who maintains the car with a low three thousand miles. *Donald Farr*

car look that appealed to the coming-of-age baby boomer generation. Introduced in April 1964, the 1965 Mustang was an immediate sales success. The majority of buyers were satisfied with the six-cylinder or base V-8s. However, for young men like Bob Gordon, Ford offered the 289 High Performance.

Over the next three years, the lines between pony car and muscle car were blurred as competitors such as the SS Camaro, HO Firebird, and even corporate cousin Cougar entered the market with big-block power. When Ford's 390 V-8 proved insufficient, Ford responded with the 428 Cobra Jet at midyear 1968 to set the stage for the Mustang and Cougar's greatest performance era.

As for Bob Gordon's 1965 289 Hi-Po Mustang, while other young men were ripping up the streets in their high-performance machinery during the 1960s, the burgundy fastback sat in a lean-to barn enclosure for nearly forty years. Son Herb eventually inherited the low-mileage Mustang and had it refurbished as a tribute to both his father and the high-performance legacy of the Hi-Po Mustang.

above: For owners like Bob Gordon who needed more than the factory 271 horsepower, Ford offered a dual-quad "Cobra" induction kit for the 289 High Performance. *Donald Farr*

opposite top: Before GTs and Shelbys, the earliest 1964½ Mustangs with the 289 Hi-Po looked very much like standard hardtops or convertibles. *Tom Shaw*

opposite bottom: When the steering column–mounted Rally-Pak was ordered, the Hi-Po's tachometer stretched to 8,000 rpm instead of 6,000.

1964½–66 Mustang 289 High Performance

At its early introduction on April 17, 1964, the 1965 Mustang offered little in the way of performance. In comments to the press gathered at the New York World's Fair, Ford general manager and lead Mustang cheerleader Lee Iacocca stressed the new Mustang's flexibility as an economy car, luxury car, or sports car. Muscle car wasn't yet part of the automotive lexicon, nor was it descriptive of the Mustang that debuted with six-cylinder or low-performance V-8 power, initially topping out with a 289 rated at 210 horsepower. In June 1964, the game changed when Ford began installing the 289 High Performance.

Already an option for the Fairlane, the 289 High Performance added street cred to the Mustang as a performance car. With solid-lifter camshaft, 595 cfm four-barrel, dual-point distributor, and streamlined cast-iron exhaust manifolds, the K-code 289 hit its 271-horsepower rating at a high-revving 6,000 rpm. A complete package, 1965 Mustangs equipped with the 289 Hi-Po also came with a Toploader four-speed, a 9-inch rear end, dual exhaust (an Arvinode system from October 1964 to March 1965), and heavy-duty suspension with quicker ratio steering and 6.95x14-inch tires. For under-hood appeal, the Mustang's Hi-Po was topped by chrome valve covers and an open-element air cleaner. Due to the high-rpm potential, air conditioning and power steering were not available.

In August 1964, a "2+2" joined the hardtop and convertible. Providing the Mustang with an even sportier look, the sleek fastback quickly became the most popular body style for the 289 High Performance in the 1965 Mustang. In April 1965, Ford celebrated the Mustang's first anniversary by offering an optional GT

1964
SU·9117
NY WORLD'S FAIR 64
MUSTANG

FUEL
OIL
AMPS
TEMP
BRAKE
FORD

Equipment Group that was ideally suited for the Hi-Po with much-needed front disc brakes and five-dial instrumentation, along with sporty visuals such as fog lights, side stripes, and through-the-valance "trumpet" exhaust tips.

The car magazines couldn't wait to get a shot at the 1965 Mustang with the 289 High Performance. *Motor Trend* hit the newsstands first with its August 1964 report after testing a well-equipped hardtop at Riverside International Raceway:

> "For a 2,980-pound car with 271 horses under the hood, performance turned out to be impressive. The car ran through our measured quarter mile in a quick 15.7 seconds, showing 89 mph on our fifth wheel speedometer as we crossed the finish line. Charging down Riverside's long back straight, the Mustang climbed to a top speed of 117 mph at 6,300 rpm. Just listening to that engine is enough to send an enthusiast into a glassy-eyed trance."

Even *Sports Car Graphic* slipped a report between its coverage of Jaguars and Fiats. Wrote technical editor Jerry Titus: "Tester Shedenhelm, used to Sprites and Minis, at first was horrified at the seemingly huge size of the Mustang. But within a very short time, he found that the optionalized Mustang handled like a reasonably small (albeit overly powerful) sporty car and could be flung through the turns with pleasure, if not abandon."

With four-seater practicality, a usable trunk, standard bucket seats, and options to order anything from a sporty economy car to a luxury cruiser, the 1965 Mustang was a huge success for Ford, selling just over 680,000 during the extended 18-month 1965 model year. According to the HiPo Mustang Registry, only 7,272 came with the K-code 289 Hi-Po.

The Mustang was on a roll when it entered 1966. Other than a mild grille update and replacement of the Falcon instrument cluster with the five-dial version from the 1965 GT, the second-year Mustang was nearly identical to the first. The 289 High Performance continued with 271 horsepower; only for 1966 it was also available with the Cruise-O-Matic transmission, which was upgraded for firmer shifts. Chirping the tires on the second and third gear upshifts was common.

opposite: Ford announced the Cruise-O-Matic option for the 289 High Performance in full-page magazine advertising. "A slushbox, it's not," said the ad copy. *Donald Farr Collection*

inset: A small "High Performance" metal plate with checkered flags placed behind the standard 289 emblem was all that identified the 289 High Performance Mustang for 1964½. *Tom Shaw*

below: Ford's 1964½ Hi-Po press car was equipped with 15-inch wheels and Firestone Super Sport tires as part of the rarely ordered and short-lived Competition Suspension option. *Archives / TEN: The Enthusiast Network Magazine, LLC*

1966 Mustang Hardtop

Okay, you can stop asking, "Why doesn't Mustang bolt Cruise-O-Matic behind the 271 solid-lifter V-8"?

Now Mustang stampedes onto the scene with Cruise-O-Matic teamed up with the 271 horsepower solid-lifter V-8 as a new option for '66.

Here, at last, is a memorable high-performance machine that you're not afraid to let your wife drive to the supermarket!

Ford's 289-cubic inch V-8 in Cobra trim was already something of a living, fire-breathing legend. The hard part was finding an automatic strong and flexible enough to handle it.

Enter Ford's lightweight 3-speed Cruise-O-Matic. Note that *3-speed*. You can match gear ratios to road conditions almost like a manual shift. Or sit back and let things shift for themselves. Cruise-O-Matic makes swift, crisp shifts you'd be glad to call your own; a slushbox it's not.

Mustang's been reaping praise since the day it came out. With this new option, it becomes automatic.

The Mustang fastback arrived in August 1964 followed by the midyear 1965 GT Equipment Group option, both providing a more sporting image for the 289 High Performance engine. *Eric English*

Once again, Mustang sales soared for 1966 with more than 607,000 produced, yet only 5,469 were equipped with the K-code 289. Of the nearly 1.3 million Mustangs sold during the 1965–66 model years, just 12,741 were powered by the 289 High Performance.

1967–68 Mustang 390

For its first two years, the Mustang frolicked mostly alone in the new pony car playground. Plymouth had countered with the Barracuda, introducing it two weeks before the Mustang in 1964, but the Mopar was little more than a fastback roof grafted onto the Valiant compact. Mustang outsold the Barracuda nearly ten to one during 1964–66.

Not surprisingly, the Mustang's success attracted competition, and it arrived in force for 1967 with Chevrolet's Camaro and Pontiac's Firebird, all available with big-block engines. In response, Ford enlarged the Mustang to shoehorn the 390 between the shock towers. "They must have fit the 390 into the Mustang with an .020-inch feeler gauge," quipped *Hot Rod*. Restyled yet retaining its Mustang personality, the 1967 model was nearly 3 inches wider, 2 inches longer, and several hundred pounds heavier—even more when equipped with the big-block.

Although the 289 High Performance was still offered, the 390 Thunderbird Special was the high-torque ticket for the 1967 Mustang. Essentially, it was the same 320-horsepower engine found in the 1967 Fairlane GT with Holley four-barrel carburetion, a hotter camshaft, and low-restriction dual exhaust. The 427 lb-ft of torque provided the Mustang with more kick off the line, but restrictive cylinder heads choked the horsepower at the top end. "The 390 engine cannot be considered hot by today's standards," reported *CARS* magazine. "However, when packed into a light Mustang, you have a pretty impressive machine."

More impressive was the 390 paired with the GT Equipment Group, which came with power front disc brakes and Special Handling Package, plus visuals such as fog lights, side stripes, and quad exhaust tips. However, for 1967 only, the GT became the GTA when ordered with Ford's new SelectShift Cruise-O-Matic transmission, which provided the driver with automatic shifting in "D" or the ability to shift manually.

Car Life tested a 390-powered fastback and came away with similar impressions as other magazines: "Once regarded as a high-performance engine, the 390 must now be thought of as a mid-range powerplant. If the increased displacement provides additional power for creature comfort, it adds little to the straight-line acceleration so prized by many enthusiasts."

CARS magazine noted the 390's potential by taking its GT fastback test car to the drag strip, where it ran a 14.95 ET with cheater slicks and traction bars. The editors noted, "What the car needs for increased performance is a little more jetting, a good set of headers, and 3.90 or 4.11 gears."

Perhaps influenced by the abundant press coverage of the Mustang's first big-block, buyers looking for performance chose the 390's torque over the Hi-Po's high-winding horsepower. Ford produced only 489 1967 Mustangs with the K-code 289 (not including Shelbys). By contrast, 29,097 were sold with the S-code 390, split almost equally between four-speeds and automatics. For the little-changed 1968 Mustang, 390 sales fell to 10,650, partially because Ford had a new big-block up its sleeve for midyear.

above: Although not as nimble as the small-block, the 390 Mustang handled well for a car with so much weight over the front wheels. *Archives / TEN: The Enthusiast Network Magazine, LLC*

top: The 390 was available in all 1967 Mustangs, not just GTs or GTAs. The 390 in a standard hardtop made for a true street sleeper. *Donald Farr*

following pages: When ordered with the new Select-Shift automatic, the 1967 Mustang GT became the GTA. *Dale Amy*

GTA

1967–68 Cougar 390

Mercury climbed aboard the pony car train for 1967 with the Cougar, a reshaped, upscale version of the Mustang with a 3-inch longer wheelbase, hideaway headlights, wide sequential-operation taillights, and Lincoln-Mercury luxury, including more insulation for a quieter ride. "For the man on his way to a Thunderbird" is how one scribe described the hardtop-only Cougar.

Like the Mustang, the 1967 Cougar's top engine was the 390, identified by Mercury as the Marauder 390 and rated at 320 horsepower, although it was the same Holley carbed big-block that had sported 335 horsepower in the 1966 Cyclone GT. Optional in any 1967 Cougar, the 390 was the standard and only powerplant for the Cougar GT, which also added dual exhaust, low-restriction air cleaner, power front disc brakes, 6-inch-wide rims with F70x14 Wide Oval tires, quicker steering ratio, and performance handling suspension. "The suspension combination produces a definitely firm ride, the sort of firmness that tells the knowledgeable driver where all four wheels are positioned at any given time," said *Car Life* in its road test. "The Cougar GT challenges the automotively inclined to seek the quick and the crooked road."

Similar to other 390 Fords, acceleration was limited by numerically low gearing. With 3.25:1 cogs and heavier weight, *Car Life's* Cougar GT clocked a 15.9-second ET at 89.1 miles per hour, albeit the test car was labored with air conditioning and the extra weight of two people and test equipment on board.

Mercury sold 12,559 Cougars with the 390 for 1967, 8,464 as standard equipment in GTs with 2,673 also equipped with the midyear XR-7 trim package that added leather seating, overhead console, and other luxury appointments. Sales for the 390 dropped to 4,855 for the little-changed 1968 Cougar.

above: The 390 was optional for the 1968 Cougar XR7-G, which added twin-snorkel hood scoop, fog lights, and styled steel wheels. The "G" stood for racer Dan Gurney, who drove Cougars in the Trans Am road-race series. *Eric English*

opposite top: The 390 Marauder big-block was optional for the 1967 Cougar, based on the Mustang but with a much more upscale appearance. *Eric English*

opposite bottom: In the standard Cougar, the 390 was equipped with an snorkel air cleaner, not the low-restriction version like the GT. *Eric English*

F
390
P

1968½ Mustang GT 428 Cobra Jet

A scoop and black stripes on the hood set the Cobra Jet–powered 1968 Mustangs apart from regular GTs. *Donald Farr*

Hot Rod called it "the fastest running Pure Stock in the history of man." And although their 13.56-second quarter-mile number came from a "not representative" 1968 Mustang fastback that had been special ordered without insulation and prepped by Holman-Moody-Stroppe, the white 428 Cobra Jet-powered fastback served notice that Ford was done playing around with passenger-car 390s. The new 428 CJ, rated at "only" 335 horsepower, propelled the Ford pony cars into supercar territory.

Ford announced the Cobra Jet in April 1968 as an option for the Mustang GT, Cougar, Fairlane, and Cyclone. But the engine had actually debuted earlier, in January 1968, at the NHRA Winternationals, where a quintet of CJ-powered fastbacks plowed through the Super Stock ranks with driver Al Joniec winning the overall Eliminator title.

The 428 Cobra Jet was the first volley in what would become a three-year supercar assault. Available as a $420 option for all three Mustang body styles with the GT Equipment Group mandatory, the Cobra Jet package added the 335-horsepower 428 with 428 Police Interceptor block, updated 427 Low-Riser heads, and 390 GT camshaft. Ford also equipped the CJ with high-flow exhaust manifolds and 735 cfm

above: The 428 Cobra Jet was available for all Mustang body styles, including the lighter-weight hardtop. Only 221 were sold.
Eric English

left: Holley carburetors had been used on performance Ford engines since the early 1960s. For the Cobra Jet, a 735 cfm version with vacuum secondaries sat on a cast-iron intake.
Donald Farr

Ram-Air

Hood scoops were not new in the 1960s, having been around since the previous decade when they served as ornamentation on the 1955 Thunderbird. The 1964 Pontiac GTO had nonfunctional scoops, while the 1965 Shelby GT350's scoop fed cooler air into the engine compartment but not directly into the engine. For the 428 Cobra Jet, Ford developed a simple contraption to "ram" cooler, thus denser, outside air into the carburetor for more power, although the increase would not be reflected in factory horsepower ratings.

First offered as standard equipment on the 1968½ CJ Mustangs and Cougars, Ford's first Ram-Air consisted of a metal plate with flapper door that sat on the air cleaner. At idle and low rpms, the engine vacuum kept the door closed, and the engine breathed through the air cleaner snorkel per normal. But when engine vacuum dropped at wide-open throttle, the door popped open to ram cooler outside air from the hood scoop. A thick, rubber gasket sealed the Ram-Air to the closed hood and also funneled rainwater to a drain tube. Instead of a full-size air cleaner lid, the Ram-Air system used a smaller-diameter lid that covered only the air filter, leaving the outer edges of the filter open. That way, air from both the scoop and snorkel were filtered before entering the carburetor.

Ford took Ram-Air to another level in 1969 with the Shaker scoop, which protruded through the hood, providing an entertaining view for the driver as the engine vibrated at idle and tilted under torque. It debuted with the Ram-Air option for the 1969 Mustang with a slightly restyled version added to the Torino for 1970. In 1971, the Mustang adopted NACA-style (Ford called them "NASA") hood ducts, made functional with a plenum, also vacuum operated, mounted under the hood.

Ford's Ram-Air for 1968½ Cobra Jet Mustangs and Cougars consisted of a scoop with its hood opening sealed to a metal plenum with a flapper door on top of the air cleaner. *Jerry Heasley*

Holley four-barrel. All 1968½ Mustang 428 CJs were R codes with Ram-Air. It was the first use of a functional hood scoop for a Mustang; it wouldn't be the last.

The 1968 Cobra Jet Mustangs also received the competition handling suspension, F70x14 white-letter Goodyear Polyglas tires, power front disc brakes, 80-amp battery, and choice of Toploader four-speed or C6 automatic. With the four-speed, the 9-inch rear end was supported by staggered shocks to reduce wheel hop during hard launches.

Ford built 1,299 Cobra Jet Mustangs for 1968 (not counting 1,571 supplied to Shelby for GT500KRs). The majority, 1,044, were fastbacks. Hardtop production reached only 221, but the rare ones were the convertibles at just 34.

The 1968½ 428 Cobra Jet was a new start for Mustang muscle; for 1969, Ford had a plan to mate the CJ to a fastback with a supercar image.

In addition to its standard 427 engine, the 1968 Cougar GT-E added European flair to the Cougar's exterior with two-tone paint, power dome hood, and blackout grille.
Jerry Heasley

1968 Cougar GT-E

The Cougar fractured its all-luxury reputation in 1968 with a new option package, the GT-E, available for $1,311 over the standard Cougar or XR-7. For the 393 buyers who plopped down the cash, the investment was worth every penny, because the Cougar GT-E was the only way to get a 427 in a 1968 Ford and Mercury.

As a limited production, race-oriented powerplant, 427s were hand-assembled, not built on an engine production line, making them more expensive to produce. According to former engine engineer Bill Barr, Ford was working on the powerplant that would become the 428 Cobra Jet, but it wouldn't be ready until mid-1968. Mercury didn't want to wait and installed the 427 as the only available engine for the 1968 Cougar GT-E.

Mercury never explained the meaning of the "E," other than sales literature implying the GT-E was "bred for excitement."

In a time when supercars were defined as stripped-down intermediates or pony cars with powerful engines, the 1968 Cougar GT-E combined big horsepower with Mercury style and luxury. However, the GT-E's W-code 427 wasn't the same as the famed solid-lifter, dual-quad big-block from the early 1960s. Rated at 390 horsepower, it was slightly detuned with lower compression, a milder hydraulic cam, and single Holley four-barrel on a Medium Riser intake. The W code also used a new cylinder head design that combined Low Riser–type intake ports with a revised exhaust port to fit the Cougar's manifolds. It would be the last 427 ever offered in a Ford or Mercury production car.

ILLINOIS
427 GTE
Mercury
Cougar
Firestone
WIDE OVAL

above: The GT-E's 7.0 Litre emblem didn't change when the 427 was replaced by the 428 Cobra Jet at midyear. *Jerry Heasley*

opposite: Blackout taillights, quad exhaust tips, and styled steel wheels were part of the 1968 Cougar GT-E package. *Jerry Heasley*

In addition to the 427, the equipment list was impressive: Super Competition Handling package, mandatory SelectShift Merc-O-Matic automatic, 9-inch rear axle, power steering, power front disc brakes, "power booster" engine fan, low-restriction dual exhaust, and chrome engine dress-up with open-element air cleaner. In contrast to its Mercury image, air conditioning and speed control were not available. Appearance-wise, the GT-E was identified by its two-tone paint with argent on the lower body, blackout grille with unique horizontal trim bar, nonfunctional twin-nostril hood scoop, black taillight trim, quad exhaust tips, styled steel wheels, and "7.0 Litre GT-E" fender emblems. *Car Life* tested an XR-7-equipped GT-E: "The 427 performed as it should have, perhaps with more refinement and better low-speed drivability than we expected. *Car Life* has never tested a more flexible engine, one which could pull strongly in top gear at 40 mph yet scream past 5,000 rpm so quickly a sharp eye on the tachometer was mandatory." In explaining the "not quite up to supercar status" 15.12-second ET, the writer said, "Bear in mind that the test car did not have any form of limited-slip rear axle. Thus, much of the 427's strong low-speed torque went up in right rear tire smoke."

The GT-E's 427 was phased out over a three-month period by the 428 Cobra Jet when it became available in the spring of 1968. All Cobra Jets were R codes with functional Ram-Air, so the twin-nostril GT-E "power dome" was replaced by a larger functional scoop and black hood stripe, as used on all 428 CJ Cougars. Essentially, the 428 displacement was close enough that the 7.0 Litre emblems remained on the fenders. A four-speed was available with the CJ, but only three were produced.

With the 427's $1,300 tab pushing the GT-E's sticker price to well over $4,000, especially when added to the XR-7, few were sold. Only 356 were produced with the 427, plus a super-rare 37 with the 428 Cobra Jet.

427 and 302 Tunnel Port

Clear as black ink on white paper, two engines were listed in the 1968 Mustang sales brochure: 427-cubic-inch Cobra V-8 with 390 horsepower and the solid-lifter 302-cubic-inch V-8 with 345 horsepower "for sedan racing, available late 1967 on special order." Both were listed as options with the Mustang's GT Equipment Group. Finally, a factory Mustang 427 and a Trans-Am small-block—with tunnel-port heads like the race cars—to compete with Chevrolet's new Z/28 Camaro.

Three months later, a new sales brochure was issued with all 427 references removed. According to Marti Auto Works's Ford database, no 1968 Mustang rolled off the factory assembly line with a W-code 427, thus handing over the 427 bragging rights to the 1968 Cougar GT-E.

Although the 302 High Performance survived the brochure revision, even receiving a D-code designation for production, it was also dead on arrival. None were produced. According to former engine engineer Bill Barr, including the 427 and 302 Tunnel Port in the sales brochure was merely an attempt to convince NASCAR and the SCCA that the special engines were being made available to the public.

Car Craft noted that the Tunnel Port small-block was "available in Mustangs and Cougars" and went into detail outlining the four-bolt mains, dual-quad Holley induction, and "tunnel-port" cylinder heads with "straight shot" ports that had pushrod tubes running through them. Ford went so far as to put together a 1968 Tunnel Port Mustang for a *Car & Driver* comparison with a Z/28. Even *C&D* seemed surprised when reviewers popped the Mustang's hood: "There it was, tunnel-port fans, right in front of our eyes. The real thing. 'Well, yes, this is your regular 12.5-to-one compression ratio, dry deck, tunnel-port 302,' allowed the Ford man. 'How many do you want?'"

C&D's 302 Tunnel Port 1968 Mustang ran a 13.96-second quarter mile and, after several laps at Lime Rock, driver Sam Posey described the handling as "like a million bucks." *C&D* summed up the Tunnel Port Mustang versus Z/28 Camaro: "Both are easily the most exciting machines we've ever driven with price tags under $10,000 and by far the best performing street cars ever."

Detuned from previous versions, the 427-cubic-inch engine listed in the 1968 Mustang sales brochure made 390 horsepower with a single Holley carb, hydraulic cam, and lower compression ratio. None were built. However, a Mercury version appeared in limited numbers in the 1968 Cougar GT-E. *Donald Farr Collection*

Car Life questioned the lack of front shock absorber rebound on its test 1968 Cougar CJ, but it sure made for great photos with the front end lifting toward the sky. *Archives / TEN: The Enthusiast Network Magazine, LLC*

1968½ Cougar 428 Cobra Jet

When the 428 Cobra Jet big-block joined the Ford powertrain lineup in April 1968, it not only replaced the 427 in the GT-E but also became a $420 option for all 1968 Cougars, including the standard hardtop, XR-7, and GT. Suddenly, the cat had claws with the CJ's underrated 335 horsepower and all the heavy-duty equipment that came with it.

Like the Mustang, all 1968½ CJ Cougars were R codes with Ram-Air, so a functional scoop was added to the hood along with a satin black stripe that extended out of the scoop and over the nose. Other than the scoop and stripe, there was no external identification to warn the driver in the next lane that the Cougar was powered by Ford's latest performance engine. Available with four-speed or automatic, the Cobra Jet Cougar was also equipped with the Competition Handling Package, 6-inch wide wheels (standard or styled steel), Goodyear F70x14 Polyglas white-letter tires, and 9-inch rear end with nodular housing and 31-spline axles. Standard gearing was 3.50:1 in an open differential. Much-needed Traction-Lok was optional for 3.50, 3.91, and 4.11 gears.

With only a few months' availability and little promotion prior to the arrival of the 1969 models, Mercury sold only 244 Cobra Jet Cougars for 1968—127 base models, 66 XR-7s, 37 GT-Es, and 14 XR7-Gs.

Most of the car magazines focused on the CJ-powered 1968 Cyclone, but *Super Stock & Drag Illustrated* scored a loaded XR-7 GT with the 428 Cobra Jet, C6 automatic (described as the "consistomatic"), and 3.91 gears for a test by Mercury drag racer Don Nicholson. With nothing more than fresh spark plugs, Nicholson's first pass netted a 14.08 ET at 100.66 miles per hour. By experimenting with burnout and launch techniques, Nicholson dropped the time to 13.81 at 101.46. Nicholson then bumped the timing (by tapping the distributer with a hammer, *Super Stock* reported) and removed the Ram-Air flapper on the air cleaner, which resulted in a 13.23 at 103.39, one of the best performances ever by a stock Ford or Mercury muscle car.

CARS magazine's Roger Huntington wrung a 14.52 at 101.92 out of a stock CJ-powered XR-7, noting, "That's right up there with the best of the supercars."

BY THE NUMBERS:

1964½ MUSTANG HARDTOP

ENGINE: 289 High Performance

CARBURETION: Autolite four-barrel

HORSEPOWER: 271 at 6,000 rpm

TORQUE: 312 at 3,400 rpm

TRANSMISSION: Four-speed

REAR AXLE RATIO: 3.89

WEIGHT: 2,980 lbs (curb)

HORSEPOWER TO WEIGHT: 10.99

QUARTER MILE: 15.7 at 89 mph (*Motor Trend*, August 1964)

1967 MUSTANG GTA 390

ENGINE: 390 Thunderbird Special

CARBURETION: Holley four-barrel

HORSEPOWER: 320 at 4,800 rpm

TORQUE: 427 at 3,200 rpm

TRANSMISSION: Automatic

REAR AXLE RATIO: 3.00:1

WEIGHT: 3,897 lbs (test)

HORSEPOWER TO WEIGHT: 12.17

QUARTER MILE: 15.2 at 91 mph (*Car & Driver*, November 1966)

1967 COUGAR GT 390

ENGINE: Marauder 390

CARBURETION: Holley four-barrel

HORSEPOWER: 320 at 4,800 rpm

TORQUE: 427 at 3,200 rpm

TRANSMISSION: Automatic

REA8 3.25:1

WEIGHT: 3,920 lbs (test)

HORSEPOWER TO WEIGHT: 12.25

QUARTER MILE: 15.9 at 89.1 mph (*Car Life*, July 1967)

1968 MUSTANG 428 COBRA JET

ENGINE: 428 Cobra Jet

CARBURETION: 735 cfm Holley four-barrel

HORSEPOWER: 335 at 5,400 rpm

TORQUE: 440 at 3,400 rpm

TRANSMISSION: Four-speed

REAR AXLE RATIO: 3.89:1

WEIGHT: 3,240 lbs (sound deadener deleted)

HORSEPOWER TO WEIGHT: 9.67

QUARTER MILE: 13.46 at 106.64 mph (*Hot Rod*, March 1968)

1968 COUGAR GT-E

ENGINE: 7.0 Litre 427

CARBURETION: Holley four-barrel

HORSEPOWER: 390 at 5,600 rpm

TORQUE: 460 at 3,200 rpm

TRANSMISSION: Automatic

REAR AXLE RATIO: 3.50:1

WEIGHT: 3,982 lbs (test)

HORSEPOWER TO WEIGHT: 10.21

QUARTER MILE: 15.12 at 96.3 mph (*Car Life*, July 1968)

1968½ COUGAR XR-7

ENGINE: 428 Cobra Jet

CARBURETION: Holley four-barrel

HORSEPOWER: 335 at 4,800 rpm

TORQUE: 445 at 3,400 rpm

TRANSMISSION: Automatic

REAR AXLE RATIO: 3.91:1

WEIGHT: 3,600 lbs (test)

HORSEPOWER TO WEIGHT: 10.74

QUARTER MILE: 13.23 at 103.39 mph (*Super Stock*, August 1968)

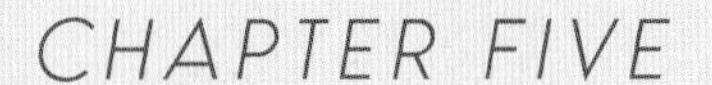

CHAPTER FIVE

PONY CARS II

1969-1973

Richard West was 19 years old and working part time at a local gas station when a striped and spoilered 1970 Boss 302 Mustang pulled in for a fill-up. "I'd never seen anything like it," he recalled. Instead of fixing up his 1965 Mustang, Richard decided to order a new Boss 302. His father drove him to McFayden's Ford Center in Omaha, Nebraska, where Richard checked off the option list—4.30 gears with Detroit Locker, Shaker hood scoop, rear spoiler, Magnum 500 wheels, tachometer, and Deluxe interior. His parents cosigned for the loan on the $4,389 Boss, less $1,100 for his 1965 Mustang trade-in. For the next three years, Richard paid $110.73 a month for the right to drive one of the hottest Mustangs on the road.

The Mustang reached its performance pinnacle with the Boss 429, rated at 375 horsepower with a NASCAR-inspired hemi-head engine. *Jerry Heasley*

Less than a year after Bob Tasca's 1967 tongue-lashing about Ford's lackadaisical approach to performance, the 1969s arrived at dealerships with new names, stripes, and spoilers, even a vibrating scoop poking through the hood. Over the next couple of years, Mustang and Cougar fans gained new words for their bench-racing vocabulary—Mach 1, Boss, Eliminator, Shaker, Super Cobra Jet, Drag Pack, and Detroit Locker. During 1969–71, Ford's pony cars were at the top of the muscle car food chain.

Throughout the 1970s, Richard West enjoyed his Boss 302. "I bought it on a Wednesday, and that weekend I was at Cornhusker Dragway," Richard said. "It only ran 14.80s that day, but I didn't know how to drive it out of the hole with 4.30 gears!" Like most young men of the era, he added the typical modifications, including headers, Cragar wheels, and air shocks so the rear fenders would clear the fat rear tires. A spun bearing in the late 1970s forced Richard to park his Mustang with less than thirty thousand miles. But unlike most, he resisted the temptation to sell and kept the Boss until he was able to restore it in the 1990s. Today, it's a reminder of his youth from a time when Mustang muscle cars were at their all-time best.

above: By 1973, Richard West's 1970 Boss 302 was the typical three-year-old muscle car with mag wheels, air shocks, and headers. *Richard West*

opposite top: With the 1969 Mach 1, Mustang finally gained a muscle image. "Speed of Sound" was fitting for a Mach with the 428 Cobra Jet and functional Shaker hood scoop. *Donald Farr*

opposite bottom left: The 1969 Mach 1 got its own interior with high-back bucket seats, console, Rim-Blow steering wheel, woodgrain trim, and red stitched-in floor mats. *Donald Farr*

opposite bottom right: In Q-code form, the Mach 1's 428 Cobra Jet engine came with a closed air cleaner. This one has the early chrome valve covers. *Tom Shaw*

1969–70 Mach 1 428 Cobra Jet

> "First gear is a rubber burning, smoking, screeching fishtail. Move the console-mounted selector lever one notch forward. Second gear slams in instantly producing more screeching. The car fishtails out to the left and you're crushed back to your seat. Move the lever once more into the Drive slot. Third comes on *right* now with a loud retort from the rear tires and you still can't move off the seat back."

That's how *Super Stock*'s Roland McGonagal described his drag test of Ford's latest Mustang supercar—the 1969 Mach 1 powered by the Ram-Air 428 Cobra Jet.

Lee Iacocca's 1967 memo inquiring, "What are we going to do about the performance image problem?" spawned a new urgency at Ford. Within months, a new 428 Cobra Jet engine was on the street in 1968 Fords. For 1969, a supercar image was stirred into the mix with the 1969 Fairlane Cobra and Cyclone CJ. And for the boldly restyled, slightly larger Mustang SportsRoof, there was the Mach 1.

In base form, the 1969 Mach 1 carried a $3,122 base price, which packaged its "speed of sound" image with blackout hood with scoop and click-pins, reflective side stripes, styled steel wheels, and quad exhaust tips for four-barrel engines. The Mach also got its own Deluxe-style interior with high-back bucket seats, woodgrain trim, console, and Rim-Blow three-spoke steering wheel, while the Competition Suspension added stiffer springs, heavy-duty shocks, and thicker front stabilizer bar. The standard two-barrel 351 Windsor kept the sticker price low but belied the Mach's supercar image; thankfully, the four-barrel 351, 390, and 428 Cobra Jet engines were available optionally. The Mach 1 earned its supercar wings when equipped with the Cobra Jet, available as a Q code or the R-code Ram-Air with the new and eye-catching Shaker

COBRA
428-4V
Autolite

UNION PACIFIC
GOODYEAR
POLYGLAS

top: Ford went with brightly colored illustrations to promote its 1970 performance Mustang lineup, which included the Mach 1. *Donald Farr Collection*

left: The 428 Cobra Jet continued as an option for the 1970 Mach 1, including a handful of *Twister Specials* built as a special Kansas promotion. They came with side decals in addition to the 1970 Mach 1's aluminum rocker panel covers. *Eric English*

scoop. Both CJs were rated at 335 horsepower and available with either automatic or four-speed, which added staggered rear shocks.

Attached to the air cleaner and poking through an opening in the hood, the Shaker attracted plenty of comments. "In a year when every manufacturer offers hood scoops, Ford outdoes them all with an AA/Fuel Dragster-style bug-catcher sticking right out through a hole in the hood," said *Car & Driver*. "Clearly, Ford has scooped the entire industry," proclaimed *Motor Trend*. "Since it's attached to the engine, it shakes and vibrates with the engine and looks tough as hell," noted *Super Stock*.

While most CJ Mach 1 drag tests clocked low 14-second passes, *Car Life* snuck into the 13s with a 3.91-geared, auto trans R code: "Previous holder of the *Car Life* passenger car record was a Plymouth Hemi with a 14-second ET," the review said. "An easy start, shifting by hand at 5,400 rpm, and the Mach holds the record—13.9 at 103. Another driver and a 13.86. Thumbing through the *Car Life* files turned up one car with a lower elapsed time, a 427 Corvette. That's a sports car."

To address overheating issues, 428 CJ Mustangs with the 3.91 or 4.30 rear axle ratios were equipped with an external oil cooler mounted in front of the radiator. Because those cars were likely to be used for racing, the engine was also strengthened with 427-style cap-screw connecting rods, which mandated revised balancing for the crankshaft, flywheel, and balance damper. These engines became known as Super Cobra Jets, although the 335-horsepower rating stayed the same. In February 1969, Ford gave the package a Drag Pack marketing name on the option list.

For 1970, the Mach 1 continued on the Mustang's updated styling—reverting to two headlights and losing the fake rear quarter panel side scoops—with new aluminum rocker panels, hood stripes, twist-type hood locks, honeycomb rear panel, and mag-look wheel covers. Mechanically, the 428 Cobra Jet was identical, including the Shaker scooped R-code engine along with Super Cobra Jet oil cooler and heavy-duty internals for cars equipped with the Drag Pack 3.91 or 4.30 gearing. Four-speed cars were updated with a T-handled Hurst shifter.

While the Mach 1 provided the image to go with the Cobra Jet name, the two 428s were optional for any 1969–70 Mustang hardtop, convertible, or SportsRoof, Ford's new name for the fastback.

The Mach 1 was a huge success, with 72,458 sold in 1969 followed by another 40,975 for 1970. However, only 15,298—12,113 for 1969 and 3,185 for 1970—backed up the supercar image with the 428 Cobra Jet.

1969–70 Cougar Cobra Jet

Compared to Mustang's muscular Mach 1, the 1969–70 Cobra Jet Cougar was a luxury sleeper. Q-code cars without Ram-Air offered no external CJ identification, while R-code versions with Ram-Air were identified only by a hood that included functional scoop, decal with "428 Cobra Jet" lettering, and click-pins.

For 1969, the two-year-old Cougar received its first makeover, retaining its basic shape but growing in both length and width. More noticeable was the sleeker grille treatment. A convertible joined the hardtop, with both body styles available as XR-7s with plush interior. The car magazines flocked to the image-rich 1969 Cyclone CJ, overlooking the

Mercury promoted the 1969 Cobra Jet Cougar as "the prowler breed for the sports car lover." The 335-horsepower engine was available in both the convertible and hardtop. *Cam Hutchins*

Mercury updated the Cougar for 1970 with a new grille design. This XR-7 is powered by the Ram-Air 428 Cobra Jet. *Eric English*

Cobra Jet Cougar except for a glossing over in the new car preview issues. *Car & Driver* said it was "for those frightened off by the Mach 1's slots and scoops."

When optioned with the 428 Cobra Jet, the 1969 Cougar was available with C6 automatic or a four-speed that also added staggered rear shocks. As with the CJ Mustangs, the 428s in Cougars with 3.91 or 4.30 gearing were upgraded to Super Cobra Jet status with oil cooler and heavier-duty bottom end. Beginning in February 1969, Mercury marketed the package as either the Drag Pak (a different spelling from Ford) with a 3.91 Traction-Lok or a Super Drag Pak with a 4.30 Detroit Locker.

Car Craft chose a 1969 Cobra Jet Cougar as a *Super Cat* project car. Calling it "a little heavier and a lot more posh" than the Mustang, staff members headed to Irwindale Raceway for a "Polyglas-melting" 14.18-second clocking at 100.55 miles per hour. By altering their launch and shift techniques, they lowered the ET to 13.93 at 102.38. The editors then delivered the Cougar to drag racer Don Nicholson, who prepped the engine and suspension for F/Stock competition. The CJ cat responded with a 12.78 at 109.35, leading the editors to ask, "What's that about 'posh luxury cars?'"

Mercury updated the 1970 Cougar with new taillights and a head-bumping hood protrusion that separated the grille. The 428 Cobra Jet remained on the option list but available only as a Q code; Ram-Air with functional hood scoop became a separate option. Mechanically, CJ-powered cars got a rear sway bar, and a Hurst shifter became standard with four-speeds.

During 1969–70, Mercury produced only 2,380 Cougars with the 428 Cobra Jet, 1,540 in 1969 and 840 for 1970.

1969–70 Boss 429

Truthfully, the NASCAR-inspired Boss 429 big-block should have gone into the aerodynamic Talladega, Ford's 1969 NASCAR warrior. Reportedly, Ford president Bunkie Knudsen supported a different direction—and maybe even suggested it. Dropping the hemi-head 429 into a Mustang would create the mother of all supercars.

Ford developed the Boss 429 engine specifically for NASCAR as a replacement for the 427, which had served Ford racing teams well since 1963 but had lost ground to Chrysler's 426 Hemi and Chevy's porcupine-head 427. Based on the new 385-series 429, the Boss big-block benefitted from aluminum heads with semi-hemispherical combustion chambers, staggered valves, and huge round ports. In NASCAR, the free-breathing Boss 429 would give the sleek-nose Ford Talladegas and Mercury Spoiler IIs a fighting chance on high-speed tracks such as Daytona and Talladega.

But before the Boss 429 could make its first lap in competition, Ford had to homologate the engine in a production car, per NASCAR's rule mandating that at least five hundred be sold to the public. The "Boss" name tied into the youthful slang of the era and was also being used to homologate a canted-valve 302 in a Trans-Am version of the Mustang. At Ford in 1969, "Boss" was the new buzzword.

For public consumption, the Boss 429 engine was detuned considerably from its competition counterpart. The short-block was strengthened with a steel crankshaft,

Ford produced the Boss 429 for two model years, 1969 and 1970, all with huge functional hood scoop, Magnum 500 wheels, and small "Boss 429" decals on the front fenders. *Jerry Heasley*

Royal Maroon was one of five colors offered for the 1969 Boss 429. Except for the hood scoop and small side decals, the Boss 429 was easily mistaken for a base Mustang SportsRoof. *Jerry Heasley*

four-bolt main bearing caps, heavy-duty rods, and forged aluminum pistons, while the cylinder heads were the same dry-deck aluminum pieces used on the race engine, sealed to the block with O-rings instead of conventional gaskets. But for polite street manners, the 1969 Boss 429 used a hydraulic camshaft, cast-iron exhaust manifolds, and aluminum intake topped by a 735 cfm Holley four-barrel. The factory rating was 375 horsepower, most ever for a Mustang to date.

With its hemi heads, the Boss 429 was too wide for the 1969 Mustang's engine compartment. When modifications needed for the installation were deemed too extensive for Ford's regular assembly line, final production was outsourced to Kar Kraft, a contracted shop that handled many of Ford's performance and prototype assignments. To accommodate the final build process, Kar Kraft acquired a former trailer manufacturing building in Brighton, Michigan. At first, Ford supplied Q-code 428 Super Cobra Jet SportsRoofs with special inner fenders that widened the engine compartment. As SCJs, these cars were also equipped with standard Boss 429 equipment, such as a four-speed, an oil cooler, a heavy-duty suspension with staggered rear shocks, and a 9-inch rear axle with 3.91 Traction-Lok differential. At Kar Kraft, the 428 drivetrain was removed and the four-speed transferred to the Boss 429 engine prior to lowering into the engine compartment. Other needed modifications were also handled at Kar Kraft, including installing larger spindles, moving the upper control arms outward, relocating the battery to the trunk, bolting on the Boss 429's large fiberglass hood scoop, and adding a ¾-inch rear sway bar, a first for a Mustang. After the initial one hundred or so cars, Ford streamlined the operation by supplying SCJ-spec SportsRoofs without engine or transmission.

Externally, the 1969 Boss 429 was available in five colors—Wimbledon White, Raven Black, Royal Maroon, Black Jade, and Candyapple Red. Other than the manually controlled hood scoop, front spoiler, small fender decals, and 15-inch Magnum 500 wheels with F60x15 Polyglas tires, the Boss 429 didn't look much different than a base six-cylinder Mustang.

Car Life was impressed: "[The Boss 429] is, quite frankly, the best enthusiast car Ford has ever produced. It ranks as one of the more impressive performance cars we've

tested. It comes standard with detail items that enthusiasts usually have to order special or add later: oil cooler, battery in the back, suspension modifications, honest spoilers. It's all there." The writer noted that the Boss 429's 14.09-second quarter-mile clocking wasn't as quick as a 428 CJ Mach 1, although the article was quick to add that the Boss "works better with open exhaust."

With its $1,208 extra cost plus mandatory options such as Interior Décor Group, power steering, and power brakes, Boss 429s stickered in the $5,000 range, a hefty price tag for a Mustang in 1969. However, demand exceeded expectations—by the end of the 1969 model year, 849 Boss 429s had rolled off Kar Kraft's assembly line, 349 more than required by NASCAR.

The Boss 429 continued into 1970 on the Mustang's slightly restyled SportsRoof. The color palette was again limited to five, but the shades were much brighter by bringing in Ford's new Grabber colors, contrasted by a black hood scoop. Mechanically, the hydraulic valvetrain was replaced by the solid-lifter cam from the 429 Super Cobra Jet, and, as with all 1970 four-speed cars, a Hurst shifter became standard equipment. Ford stuck with the NASCAR numbers for 1970, building only the mandated five hundred to bring the total two-year Boss 429 production to 1,349.

Ford pulled out of racing in August 1970, thus eliminating the need for a special racing engine and ending the short-lived era of the Boss 429 Mustang.

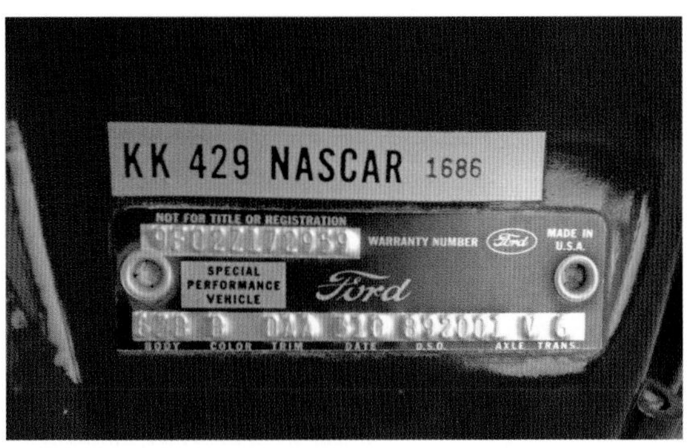

above: Every Boss 429 came with a special door tag with its KK 429 NASCAR number, plus "Special Performance Vehicle" on the door data plate. *Jerry Heasley*

top: Because installation of the hemi-head engine was too complicated for the factory assembly plant, the Boss 429s were completed on an assembly line at Kar Kraft, Ford's contracted performance company. *Donald Farr Collection*

top: For 1970, the Boss 429 was again available in five colors, only different, to include Ford's new Grabber shades, such as this one in Grabber Orange. The hood scoop was also painted black for 1970. *Jerry Heasley*

bottom: Like all 1970 four-speed Mustangs, the Boss 429's shifter was upgraded to a T-handle Hurst. Deluxe woodgrain trim was included, except the door panels. The original owner added the Boss 429 plaque and oil pressure gauge. *Jerry Heasley*

1969–70 Boss 302

The SCCA was not pleased with Ford after 1968, a season when the competition Trans-Am Mustangs were powered by Tunnel Port 302s, which were never installed in production street cars as required by the one thousand–car minimum rule. Ford rectified the situation for 1969. When the SCCA sent a tech inspector to Dearborn, Michigan, prior to the start of the Trans-Am season, he found Boss 302 Mustangs pouring off the assembly line.

Introduced in April 1969, the Boss 302 gave Ford a legitimate Trans-Am Mustang, one that would also provide an antidote to the Chevrolet's popular 302-powered Z/28 Camaro. The Boss engine was Ford's latest for Trans-Am: a four-bolt-main 302 with canted-valve, large-port heads borrowed from 1970's upcoming 351 Cleveland four-barrel. Minor modifications were required to adapt the Cleveland heads to the Windsor block, but because the heads were destined for production, they were less expensive to produce than Tunnel Ports. In production form with a solid-lifter cam, Holley 780 cfm four-barrel on an aluminum intake, high-flow exhaust manifolds, a dual-point distributor, and a 10.5:1 compression ratio, the street Boss 302 was rated at 290 horsepower, same as the Z/28.

Ford's 1969 NASCAR drivers were happy to pose with the aerodynamic Talledega powered by the Boss 429.

The small-block Boss came with Ford's best heavy-duty equipment: Toploader four-speed, nine-inch axle, special handling suspension, staggered rear shocks, and 15x7-inch Magnum 500 wheels with Goodyear's new F60 Polyglas tires, which forced engineers to strengthen the shock towers with additional bracing. The Boss 302 stood out from the supercar crowd with reflective C striping, a front spoiler, and blackout for the headlight buckets and rear panel, plus a Mach 1-like black hood. Colors were limited to Bright Yellow, Calypso Coral, Acapulco Blue, and Wimbledon White. Ford designer Larry Shinoda also eliminated the faux side scoops in the SportsRoof's quarter panels, explaining that "scoops should scoop something, and since those didn't, they had to go." Shinoda also gets credit for the Boss name, recommending it as "something the kids can relate to."

above: With canted-valve Cleveland heads, solid-lifters, and a Holley 780 cfm four-barrel on an aluminum intake, the Boss 302 developed 290 horsepower at 5,800 rpm. *Donald Farr*

top: Ford built the street Boss 302 to homologate the canted-valve engine for Trans-Am racing, which resulted in a 1970 championship. *Donald Farr Collection*

above: Rear spoiler and Sport Slats were popular options for the Boss 302, seen here as the restyled 1970 model with designer Larry Shinoda's wild striping. *Jerry Heasley*

left: The Shaker hood scoop was added to the Boss 302's option list for 1970. Flat-top finned aluminum valve covers were new for the second year. *Jerry Heasley*

right: *Hot Rod* tested an early production 1970 Boss 302, clocking a 14.62 quarter-mile at 97.50 miles per hour and describing Ford's Trans Am pony car as "A Boss to Like." *Archives / TEN: The Enthusiast Network Magazine, LLC*

Two new options for the 1969 Boss 302—a pedestal-mount rear spoiler and Sport Slats rear window louvers—would become popular visual options and add-on accessories for Mustang fastbacks. Both were designed by Shinoda.

Along with the Z/28, the Boss 302 was a new kind of muscle car. Unlike the typical supercar of the era with a nose-heavy big engine, the Trans-Am pony cars could also handle.

Car & Driver beat the other magazines to the Boss 302 by driving an early production car on Ford's test track, a review proclaiming, "The Boss 302 is another kind of Mustang. It simply drives around the turns with a kind of detachment never before experienced in a street car wearing Ford emblems. It's easily the best Mustang yet."

The Boss 302 package added nearly $700 to the cost of the 1969 Mustang SportsRoof. In spite of its late introduction, Ford sold 1,628, well over the 1,000 minimum required by the SCCA for Trans-Am.

Stripes were a staple 1960s design element for American performance cars, but the design penned by Shinoda for the 1970 Boss 302 took striping to a new extreme, running from the leading edge of the hood, down over the sides, and all the way back to the rear bumper. "Externally, there are enough markings on this car to magnetize friendly lawmen right off their freeway on-ramp perches and hold their attention for many nervous miles," said *Hot Rod*.

In addition to the new model year styling updates, the 1970 Boss 302 was available in any Mustang color, including the new Grabber shades, and came with plain 15-inch wheel covers and trim rings (with Sporty covers or Magnum 500s optional), plus the newly added Hurst shifter and rear sway bar. The functional Shaker hood scoop was also added as an option. To assist engine cooling, an oil cooler was installed on cars with 3.91 or 4.30 gears, although it was not marketed as a Drag Pack as with 428 Cobra Jets.

The 1969 Cougar Eliminator could back up its performance image for either drag racing or road course with available 428 Cobra Jet or Boss 302 power. *Jerry Heasley*

1969–70 Cougar Eliminator

Cougar joined the supercar ranks in the spring of 1969 with the debut of the Eliminator, a name that had appeared previously on a Cougar concept car. The word came from drag racing "eliminations," plus it tied in with Mercury's "Streep Scene" marketing. With a drag-race name and spoilers inspired by road racing, the Eliminator was an image conflation. Thankfully, the latest Cougar could back up both with the availability of the road racing-inspired Boss 302 and the drag-strip appropriate 428 Cobra Jet.

The Eliminator was an all-around performance package. Visually, it was the wildest Cougar yet with hood scoop, front and rear spoilers, blackout grille, argent-styled steel wheels, and a black or white mid-body stripe on one of four available colors—Competition Orange, Yellow, Bright Blue, or White. Inside, the standard Cougar hardtop interior was upgraded with high-back bucket seats, unique instrument panel with tachometer, Visual Check Panel, elapsed time clock, and Rim-Blow steering wheel. Mechanically, the Competition Suspension added heavy-duty shocks, stiffer springs, and a larger front sway bar.

Unlike Mustang's Boss 302, the 1969 Eliminator was available with a variety of powerplants, starting with the standard 351 Windsor four-barrel and including the optional 390 four-barrel and 428 Cobra Jet, with the Boss 302 added at midyear 1969. Transmission choices included three-speed (351 and 390), automatic (except Boss 302), and four-speed.

While the 290-horsepower 351 and 320-horsepower 390 were no slouches, the Boss 302 and 428 Cobra Jet put elimination possibilities into the Eliminator. Although rated at the same 290 horsepower as the four-barrel 351, the Boss 302

was a true high-performance powerplant backed by mandatory four-speed and nine-inch rear axle with staggered shocks. Likewise, Eliminators equipped with the 335-horsepower 428 Cobra Jet were packaged with heavy-duty equipment, including staggered shocks for four-speed cars. The CJ was available as either the Q code or the R code with functional Ram-Air. Either way, the Eliminator's hood was topped by a scoop but functional only with the R-code 428.

Mercury sold 2,250 Cougar Eliminators over its half-year 1969 production span. Nearly 70 percent were ordered with the base 351 V-8. Only 169 came with the Boss 302 and another 302 vehicles came with the 428 Cobra Jet, 59 as Q codes and 243 with the R code.

The Cougar Eliminator continued into 1970 with a handful of changes, including the new model year's revised front and rear styling. Eliminator specific updates included black hood scoop, hood stripe, upper body stripes with "Eliminator" lettering on the rear quarter panels, and a stripe on top of the rear spoiler with "Eliminator" identification. Exterior colors were expanded to six, including the new Competition shades of Orange, Gold, Blue, Yellow, and Green, most matching Ford's Grabber colors. Under the hood, the new 300-horsepower 351 Cleveland four-barrel replaced the 351 Windsor. The Boss 302 remained on the option list, as did the 428 Cobra Jet that was offered only as the Q code with Ram-Air available as a separate option. A rear sway bar was also added to the Boss 302 and 428 Cobra Jet suspensions.

above: With Ram-Air 428 Cobra Jet power, the Eliminator could live up to its name on the drag strip. *Jerry Heasley*

top: A fresh grille treatment and new striping differentiated the 1970 Eliminator from the previous year. *Jerry Heasley*

Mercury's 1970 Cougar advertising included the Eliminator: "Spoilers hold it down. Nothing holds it back." *Donald Farr Collection*

For 1970 Cobra Jets, Mercury marketed the optional 3.91 and 4.30 gears in a pair of packages: Drag Pak with 3.91 Traction-Lok or Super Drag Pak with 4.30 Detroit Locker, both adding an external oil cooler and strengthened bottom end to create a 428 Super Cobra Jet. The Super Drag Pak was also available for Boss 302-powered Eliminators but with the oil cooler as the only engine update.

The 1970s were out before *Car Life* could acquire a Super Drag Pak Boss 302 Eliminator for testing. *Car Life* described the 4.30:1 Detroit Locker experience as

> ". . . more clatter than a Bahamas steel drum band. The mechanical lifters and dual exhaust joined in so you have a heck of a symphony of sound. A first-gear cruise at 30 mph made more noise than you could get by tossing an oil drum down the steps of the Washington Monument."

Noting that the Cougar's extra insulation and sheet metal added 400 pounds over the Boss 302 Mustang, the *Car Life* testers flogged the 4.30-geared Eliminator to high 15-second ETs. "The low gearing put us awfully near the redline unless we stayed in Fourth gear. If ever a car needed a five-speed, this car did."

For 1970, Mercury added another 2,268 Eliminators to the two-year tally, including 469 with the Boss 302 and 247 with the 428 Cobra Jet.

1971 Mach 1 429 Cobra Jet

All of the excesses of the muscle car era came crashing into the 1971 Mustang Mach 1. Still based on the SportsRoof, the Mach 1 grew in every dimension—overall length by 2.1 inches with an inch longer wheelbase, width by 2.4 inches, and 3-inch wider front tread width. Yet the new "flatback" roofline gave the Mach a sleeker appearance, one that was backed up by the availability of the new and more powerful 429 Cobra Jet.

The 385-series 429 had earned its performance reputation when offered as a Cobra Jet in the 1970 Torino and Montego. The 1971 Mustang's larger size, including the engine compartment, accommodated the canted-valve 429's extra width compared to the previous year's FE 428. Rated at 370 horsepower, the 429 CJ boasted thirty-five more ponies than the 428 CJ, a good thing because the 1971 Mustang was also several hundred pounds heavier than the 1970 model.

above: As with 1970 models, four-speed 1971 Mustangs came with a Hurst shifter. An optional Mach 1 Sports Interior added special two-tone bucket seats. *Jerry Heasley*

top: Even though it was the largest Mustang yet, the 1971 SportsRoof styling was lean and mean, especially with the Mach 1's two-tone paint treatment and optional "hockey-stick" side stripes. *Eric English*

As with 1969–70, the 1971 Mach 1 was the performance image Mustang. The $3,268 entry-level Mach was powered by a 302 two-barrel, but paying extra supplied more power with a two- or four-barrel 351 Cleveland or one of two 429 Cobra Jets, C-code CJ or J-code 429 CJ-R with functional Ram-Air. Mach style included a honeycomb grille with sport lamps, matching honeycomb rear panel, color-keyed urethane front bumper, black or argent lower-body side paint, flat hubcaps with trim rings, and Mach 1 front fender and trunk decals with "hockey stick" side stripes optional. New twin hood ducts—the so-called "NASA style"—were standard with certain engine packages and functional only when equipped with Ram-Air, which added a plenum to the underside of the hood to funnel cooler outside air into the air cleaner. The Mach 1's special interior fell off the standard equipment list; instead, a snazzy Mach 1 Sports interior was offered at extra cost.

For all-out Mach muscle, the 429 Cobra Jet was the way to go. With 370 horsepower and 450 lb-ft of torque, the new canted-valve big-block was capable of handling air conditioning and power steering while also running low 14-second quarter miles. Both 429s were available with C6 automatic or Hurst-shifted Toploader four-speed with standard 3.25:1 non-locking differential in the 9-inch rear axle housing.

The true 429 barnstormer came by ordering the optional Drag Pack, either 3.91 Traction-Lok or 4.11 Detroit Locker. Unlike the Drag Pack 428 from 1969–70, the option did not include an oil cooler, although a kit was offered through Ford for dealer or owner installation. However, the enhancements that took the 429 CJ to Super Cobra Jet status went much further by supplying forged pistons, solid-lifter cam, and a four-barrel 780 cfm Holley. Undoubtedly, the hotter cam and larger carb added more than the five horsepower—up to 375—granted by Ford's factory rating.

The car magazines focused on the Boss 351 and barely touched on the 429 Mach 1 in their new-for-1971 stories, with most satisfied to settle for a drive at Ford's test grounds. *Motor Trend* clocked a 14.61-second quarter mile in a CJ-R Mach 1 equipped with automatic transmission, 3.25 gearing, and air conditioning. "It is a decent mixture for those who want good performance and some comfort," the review said.

In 1971, Ford sold 1,865 Mustangs with the 429 Cobra Jet, most in Mach 1s but a few as regular SportsRoofs, convertibles, and hardtops—even a handful in the luxury Grande. Only 613 were Super Cobra Jets. For 1972, the 429 CJ would disappear from the Mustang's option list, making it the last of the big-block Mustangs.

The real sleeper of 1971 was the 429 Cobra Jet in the Mustang coupe. Both versions of the 429 CJ—with both available as SCJs—were optional in all 1971 Mustangs. *Jerry Heasley*

above: When equipped with the Drag Pack, the 429 Cobra Jet was transformed into a 375-horsepower Super Cobra Jet with Holley carb, solid-lifters, and forged pistons.
Eric English

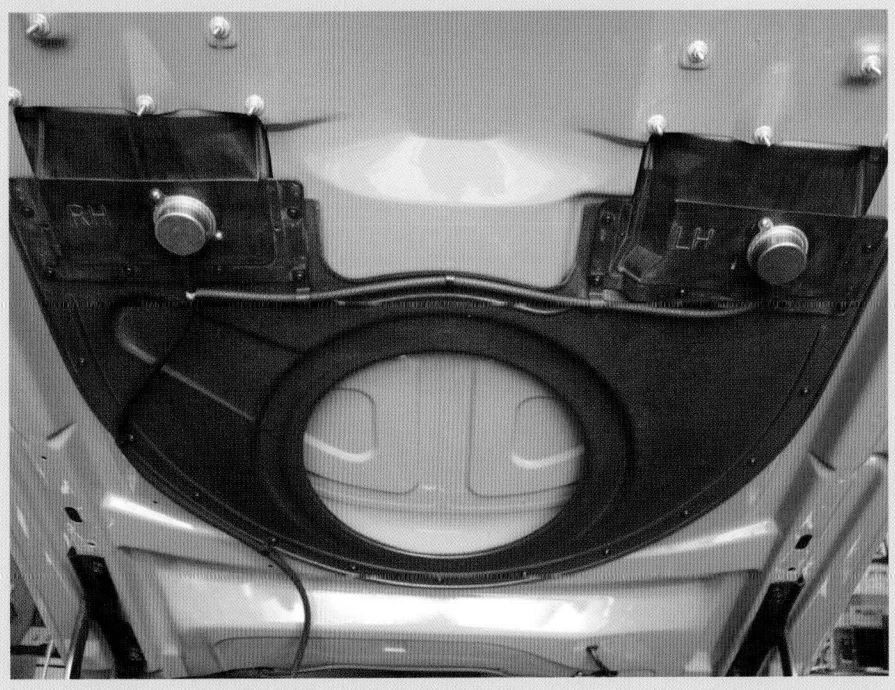

right: With the J code's Ram-Air, a plenum mounted under the hood funneled outside air from the open hood ducts to the air cleaner. Like previous Mustang Ram-Air systems, internal flaps closed and opened via engine vacuum. *Donald Farr*

The hood scoop was all that gave away the under-hood power of the 1971 Cougar with the Ram-Air 429 Cobra Jet. This XR-7 is one of only forty-seven convertibles with the 429. *Jerry Heasley*

1971 Cougar 429 Cobra Jet

Like the Mustang, the 1971 Cougar grew in all dimensions. But while the Mustang continued with a sporty look, the Cougar took on a more formal appearance, starting with the front end, which for the first time did not incorporate hideaway headlights. The larger body, longer wheelbase, and new suspension played right into the Cougar's luxury wheelhouse. Found in the mix, however, was the 429 Cobra Jet, made possible by the 1971 Cougar's larger engine compartment.

Available for the standard Cougar, XR-7, and return of the GT hardtop, the 429 CJ was offered as a C code with flat hood or J code with functional hood scoop. Both were rated at 370 horsepower and came with either the C6 automatic or four-speed with Hurst shifter. Rear axle ratios were limited to 3.25 (mandatory with air conditioning) or 3.50, with Traction-Lok optional for both. The digger-geared Drag Pak and Super Drag Pak options from 1970 were no longer offered, so the 429 Super Cobra Jet never found its way into the 1971 Cougar.

Cougars powered by the 429 Cobra Jet were also equipped with Competition Suspension, including staggered rear shocks for both automatics and four-speeds. Other mandatory equipment included dual exhaust, power brakes, larger radiator,

A Hurst four-speed shifter looks out of place in the lap of XR-7 luxury, but it happened in the final year of all-out Cobra Jet performance for the Cougar. *Jerry Heasley*

and heavy-duty battery. Standard wheels were 14x7 with F70 tires; 15-inch raised-white-letter F60 tires were optional on 15x7 wheels with hubcaps and trim rings.

CARS magazine tested a 1971 Cougar GT with the 429 Cobra Jet Ram-Air but killed the story when it wasn't able to get it to press before the 1972 models were introduced. However, writer Scott Stevens revisited his notes a decade later for a story in *Muscle Cars*:

> "It was immediately apparent that there was no way the F70x14 tires were going to harness the 429's massive torque. It was virtually impossible to launch with less than a hundred feet of wheel spin and more smoke than you'll find in an oil refinery. So we eased out as best we could, then buried our right foot. The best time slip read 14.65 seconds at 98 mph in the quarter mile. We wished we could have bolted on a set of slicks to find out how the big cat would have run with some traction. Our guess is high 13s."

With emphasis on luxury instead of performance, only 448 1971 Cougars were produced with the 429 Cobra Jet, 401 hardtops and 47 convertibles.

1971 Boss 351

Ford pulled out of racing in August 1970, eliminating the need to produce a special Boss 302 and opening the door for the larger displacement Boss 351. No longer restricted to under 5.0 liters for homologation purposes, Ford put the Boss 302 out to pasture and put 49 more cubic inches under the large-port Cleveland heads for more power, more torque, and an overall improved driving experience. Slipped into the Mustang lineup in November 1970 and packaged into the larger but sleeker-looking 1971 SportsRoof, the Boss 351 was the complete muscle car package in what would be the final year for all out Mustang performance.

The 351 Cleveland had found its way into the Mustang for 1970 in both two-barrel and four-barrel form. The 300-horsepower 4V not only featured two additional carburetor venturi, but the cylinder heads also boasted considerably larger ports

and valves. As used on the 1969–70 Boss 302, the heads were right at home on the Cleveland-based Boss 351 with its larger displacement, solid-lifter cam, and 750 cfm Autolite four-barrel on aluminum intake. Underneath, the short-block was plenty stout with four bolts at all five main bearing caps, nodular iron crank, and forged aluminum pistons on forged connecting rods.

The Boss 351 came with all the goods, including Ram-Air, mandatory four-speed with Hurst shifter, and 31-spline 9-inch rear axle supported by staggered shocks. Other performance Mustangs offered a choice of gear ratios and differentials, but the Boss 351 was supplied only with the 3.91 Traction-Lok. Exclusively, the Boss 351's Competition Suspension added 15-inch wheels, either standard steel rims with flat hubcaps or optional chrome Magnum 500s.

From a distance, the Boss 351's hockey-stick side decals and lower body/hood paint, along with black grille with driving lamps, made it easily confused with the Mach 1. However, for the Boss 351, the hood paint (black or argent, depending on exterior color) covered the majority of the NASA-scooped hood and did not taper toward the front like the Mach. Closer inspection showed that the block lettering on the fenders and rear

With its mid-displacement, canted valve engine, and other equipment, the 1971 Boss 351 was one of the best all-around performance Mustangs of the muscle-car era. *Dale Amy*

deck spelled out "Boss 351." Underneath the chrome bumper (unlike the Mach's color-keyed urethane), a newly designed front spoiler was part of the package, with rear spoiler optional.

While the car magazines complained about the 1971 Mustang's bulk and rearward visibility through the SportsRoof's "mail slot" rear window, they raved about the latest Boss Mustang's newfound horsepower and torque. "O dear mother, this thing goes like hell!" said *Sports Car Graphic*. "[The Boss 351] offers drag strip performance that most supercars with 100 cubic inches more displacement will envy," noted *Car & Driver*. Most drag tests reported low 14-second ETs. However, *Drag Racing USA* used a bone-stock Boss 351 as part of a "Mr. 4 Speed" competition between four well-known professional drag racers. In the capable hands of Ronnie Sox, the Boss 351 ran a best of 13.67 at 106 miles per hour, leading the editors to describe it as "one of the real street sleepers of the year."

With little fanfare other than a few magazine reports as Ford transitioned from performance to emissions, Boss 351 sales tallied only 1,805.

above: With its two-tone paint and hockey-stick stripes, the Boss 351 resembled the Mach 1. These decals on the front fenders set everyone straight. *Donald Farr*

top: *Hot Rod*'s Boss 351 test resulted in an impressive 14.09 quarter-mile time. "The Boss 351 is going to salt away quite a few Z/28s before this season is up," it said. *Archives / TEN: The Enthusiast Network Magazine, LLC*

1972 Mustang 351 HO

While the 1972 Mustangs were essentially unchanged on the outside, the bad news came from the inside—both the Boss 351 and 429 Cobra Jet were discontinued, victims of federal emissions and safety regulations as Ford and the other US auto manufacturers switched gears from horsepower to lower compression ratios, tamer camshafts, and

leaner fuel enrichment. Horsepower ratings took a hit, made to look even worse by 1972's new SAE net standard for measuring engine output.

But Ford wasn't quite ready to throw in the Mustang's performance towel. Quietly, in February 1972, Ford added an R-code 351 High Output, essentially a detuned Boss 351. A Ford memo justified the HO: "Although a low-volume option and special merchandising and promotion are not believed warranted in light of other priorities and limited budgets, the 351 HO is believed to be an extremely attractive product to competition-oriented customers and offers potential image value to the entire Mustang lineup."

Rated at 275 net horsepower, the 351 HO used the Boss 351's four-bolt main block and canted valve heads, but the updated engine was emissions compliant with a low 8.8:1 compression ratio, less radical solid-lifter camshaft, and adjusted ignition advance and carburetor settings. Ram-Air was not available. Not limited to a special model like 1971, the 351 HO was offered for all 1972 Mustangs, with the package also supplying Boss-like equipment, including flat-top aluminum valve covers, mandatory four-speed with Hurst shifter, 3.91-geared 9-inch rear end, power front disc brakes, and Competition Suspension with F60x15 tires on 7-inch rims.

Car & Driver made a trip to Ford's proving grounds to test the HO-equipped Mustang. "The 351 HO is no Little Lord Fauntleroy out to play," it reported. "It's big and self-assured. There is little demand for finesse, only machismo." Quarter-mile testing showed that the 351 HO was only slightly off the Boss 351's pace.

above: Unlike the 1971 Boss 351, the Boss-like 351 HO was available in all 1972 Mustangs. The majority of HOs were ordered for Mach 1s. *Donald Farr*

opposite top: Although the 351 4V was available in all 1972–73 Mustangs, it was best packaged into the Mach 1, seen here as a 1973 model. The Mach added blackout grille, side stripes, and Competition Suspension. *Jerry Heasley*

opposite bottom: Four-speed 1973 Mustangs were still supplied with a Hurst shifter. An available Instrumentation Group added tachometer and triple gauges in the center console stack. It was standard with the Mach 1 Sports Interior, seen here. *Jerry Heasley*

With late introduction and no promotion, Ford sold only 398 1972 Mustangs with the 351 HO engine. The majority were Mach 1s, but the HO was also installed in a few standard Mustangs—32 SportsRoofs, 19 hardtops, and 13 convertibles.

1972–73 Mach 1 351 4V

While the late-1972 351 HO provided the Mustang with one last performance gasp for 1972, the 351 4V was its workhorse stablemate. Compared to Boss and 428 Mustangs from just two years earlier, the Q-code Cleveland (often referred to as the "351 Cobra Jet") lost some punch due to lower compression and retarded ignition timing. However, ranked alongside other performance cars of 1972–73, the 351 4V provided the Mustang with impressive acceleration.

Ford may have been touting emissions compliance and fuel mileage in 1973, but performance vestiges such as the rear spoiler were still offered for the Mach 1 SportsRoof. *Jerry Heasley*

Rated at 280 horsepower when it debuted near the end of 1971 production, the Q-code 351 continued as an option for all 1972 Mustangs at a lower net-rated 266 horsepower with either automatic or four-speed with Hurst shifter. The four-barrel cylinder heads boasted the same huge ports and valves as before, but the switch to an open chamber design lowered both compression ratio and emissions. The 351 4V also used a hydraulic-lifter cam, cast-iron intake with Autolite four-barrel, and revised timing calibrations, including dual-point distributor in four-speed cars. In a sign of the times, the horsepower rating wasn't even mentioned in the sales brochure.

The Q-code 351 4V was ordered for 10,249 1972 Mustangs—hardtop, convertible, and SportsRoof, even the luxury Grande. However, it was at its best when found under the hood of the Mach 1, still offered with Competition Suspension and image exterior components like the 1971s. Popular options included the Mach 1 Sports interior, Magnum 500 wheels, hockey-stick side stripes, and NASA duct hood with argent or black paint and twist-type hood locks. The hood openings were not functional with the 351 4V; strangely, Ram-Air was available only for the two-barrel 351 during 1972–73.

Traditionally, the Mustang had been restyled every two years. But for 1973, the fourth-generation Mustang continued into its third year with little change other than front and rear updates, including larger "energy absorbing" front bumpers to adapt to new crash standards. The Q-code 351 4V remained as the top engine option with automatic or Hurst-shifted four-speed. With rumors flying that the Mustang would be downsized for 1974 with no V-8 offered, Q-code sales increased to 12,558 for 1973.

Road Test magazine reported a 16.2-second quarter from a four-barrel 351-powered 1973 Mach 1, noting, "Comparing the performance of this engine/transmission combination with that of previous Mustangs is meaningless due to the ever-changing emissions requirements. Suffice to say that the Mach 1 has adequate power for most of today's motoring needs."

BY THE NUMBERS:

1969 MACH 1 428 COBRA JET

ENGINE: 428 Cobra Jet Ram-Air

CARBURETION: Holley four-barrel

HORSEPOWER: 335 at 5,200 rpm

TORQUE: 440 at 3,400 rpm

TRANSMISSION: Automatic

REAR AXLE RATIO: 3.91:1 Traction-Lok

WEIGHT: 3,715 lbs

HORSEPOWER TO WEIGHT: 11.08

QUARTER MILE: 13.86 at 103.32 mph (*Car Life*, March 1969)

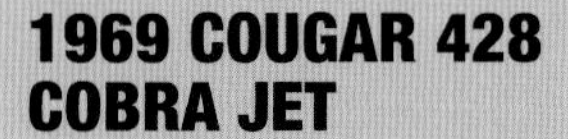

1969 COUGAR 428 COBRA JET

ENGINE: 428 Cobra Jet

CARBURETION: Holley four-barrel

HORSEPOWER: 335 at 5,200 rpm

TORQUE: 440 at 3,400 rpm

TRANSMISSION: Four-speed

REAR AXLE RATIO: Unknown

WEIGHT: Unknown

HORSEPOWER TO WEIGHT: N/A

QUARTER MILE: 13.93 at 102.38 mph (*Car Craft*, April 1969)

1969 BOSS 429

ENGINE: Boss 429

CARBURETION: 735 cfm Holley four-barrel

HORSEPOWER: 375 at 5,200 rpm

TORQUE: 410 at 3,400 rpm

TRANSMISSION: Four-speed

REAR AXLE RATIO: 3.91:1 Traction-Lok

WEIGHT: 3,870 lbs (test)

HORSEPOWER TO WEIGHT: 10.32

QUARTER MILE: 14.09 at 102.85 mph (*Car Life*, July 1969)

BY THE NUMBERS:

1969 BOSS 302

ENGINE: Boss 302

CARBURETION: 780 cfm Holley four-barrel

HORSEPOWER: 290 at 5,800 rpm

TORQUE: 290 at 4,300 rpm

TRANSMISSION: Four-speed

REAR AXLE RATIO: 3.50:1

WEIGHT: 3,485 lbs (test)

HORSEPOWER TO WEIGHT: 12.01

QUARTER MILE: 14.62 at 97.50 mph (*Hot Rod*, January 1970)

1970 COUGAR 302 ELIMINATOR

ENGINE: Boss 302

CARBURETION: Holley 780 cfm four-barrel

HORSEPOWER: 290 at 5,800 rpm

TORQUE: 290 at 4,300 rpm

TRANSMISSION: Four-speed

REAR AXLE RATIO: 4.30 Detroit Locker

WEIGHT: 4,040 lbs (test)

HORSEPOWER TO WEIGHT: 13.93

QUARTER MILE: 15.8 at 90 mph (*Car Life*, April 1970)

TRANSMISSION: Automatic

REAR AXLE RATIO: 3.25:1

1971 COUGAR GT 429

ENGINE: 429 Cobra Jet Ram-Air

CARBURETION: 715 cfm Quadrajet four-barrel

HORSEPOWER: 370 at 5,400 rpm

TORQUE: 450 at 3,400 rpm

TRANSMISSION: Automatic

REAR AXLE RATIO: 3.25:1

WEIGHT: 3,842 lbs (test)

HORSEPOWER TO WEIGHT: 10.38

QUARTER MILE: 14.65 at 98.0 mph (*MuscleCars*, via *CARS*)

1971 MACH 1

ENGINE: 429 Cobra Jet Ram-Air

CARBURETION: 715 cfm Quadrajet four-barrel

HORSEPOWER: 370 at 5,400 rpm

TORQUE: 450 at 3,400 rpm

WEIGHT: 3,805 lbs (curb)

HORSEPOWER TO WEIGHT: 10.28

QUARTER MILE: 14.61 at 96.80 mph (*Motor Trend*, January 1971)

1971 BOSS 351

ENGINE: Boss 351

CARBURETION: 750 cfm Autolite four-barrel

HORSEPOWER: 330 at 5,400 rpm

TORQUE: 370 at 4,000 rpm

TRANSMISSION: Four-speed

REAR AXLE RATIO: 3.91

WEIGHT: 3,560 lbs (curb)

HORSEPOWER TO WEIGHT: 10.78

QUARTER MILE: 14.1 at 100.6 mph (*Car & Driver*, February 1971)

1972 MUSTANG 351 HO

ENGINE: 351 Cleveland HO

CARBURETION: Autolite four-barrel

HORSEPOWER: 275 at 6,000 rpm

TORQUE: 286 at 3,800 rpm

TRANSMISSION: Four-speed

REAR AXLE RATIO: 3.91:1 Traction-Lok

WEIGHT: 3,604 lbs (curb)

HORSEPOWER TO WEIGHT: 13.10

QUARTER MILE: 15.1 at 95.6 mph (*Car & Driver*, March 1972)

1973 MACH 1 351 4VCJ

ENGINE: 351 Cleveland 4vCJ

CARBURETION: Single four-barrel

HORSEPOWER: 266 at 5,400 rpm (SAE rating)

TORQUE: 301 at 3,600 rpm

TRANSMISSION: Automatic

REAR AXLE RATIO: 3.25

WEIGHT: 3,680 lbs

HORSEPOWER TO WEIGHT: 13.83

QUARTER MILE: 16.2 at 88.7 mph (*Road Test*, July 1973)

G.T. 350
GOODYEAR

CHAPTER SIX

SHELBY

1963-1970

Hunt Palmer-Ball was nineteen when he made a quick U-turn to get a closer look at the Mustang in Louisville's Girder Motors showroom. "They had closed for the day," he remembered of that late afternoon in 1967. "I shielded my eyes with my hands and pressed my nose to the glass so I could get a better look. It was the coolest car I'd ever seen!"

Palmer-Ball had discovered the 1967 Shelby, a Mustang so different that it barely resembled the Mustang fastback it was based on. With a good-paying job at Burns Ford on the other side of town, Palmer-Ball convinced his father to cosign for a loan and placed an order for a dark blue GT350. "I put down $850 and financed the balance for three years," Palmer-Ball said. "My payments were $160 a month."

By 1960, Carroll Shelby had been driving race cars for nearly a decade and making a name for himself in both US and European road-race circles. *Sports Illustrated* bestowed him with its "Driver of the Year" honor in 1956 and 1957, and his career continued to soar when he won the 24 Hours of LeMans as an Aston Martin co-driver in 1959. But chest pains led to a serious diagnosis; Shelby suffered from a heart ailment. Doctors recommended hanging up his helmet. Shelby drove his final 1960 races with nitroglycerin tablets under his tongue.

Hunt Palmer-Ball bought his 1967 GT350 brand-new. He restored the car with only 8,600 miles on the odometer, most of them a quarter-mile at a time. *Juan Lopez-Bonilla*

However, Shelby used his bad luck as an opportunity to chase his dream of building his own sports car, one based on a European chassis but with the power of an American V-8. Through his industry contacts, Shelby learned two things in 1961: England's AC Ace was losing its engine supplier, and Ford was introducing a compact and lightweight 221-cubic-inch V-8. Shelby managed to convince both companies to cooperate with his quest.

Shelby claimed that the name came to him in a dream. Waking up one morning, he said he found "Cobra" scrawled on the notepad beside his bed.

Shelby's timing was impeccable. When his Ford-powered Cobras started winning races and beating Corvettes, Shelby found himself at the leading edge of Ford's Total Performance campaign. Similar to Holman-Moody for NASCAR, Shelby American became Ford's go-to company for sports car racing. While street Cobras garnered positive publicity from car magazines, the racing Cobras racked up championships. Soon, Ford asked for assistance with its Ford GT program. Shelby took over in 1965; one year later, Shelby American's GT40s finished one-two at the 1966 24 Hours of Le Mans. Just four years after dropping a Ford engine into his first Cobra, Carroll Shelby was on top of the racing world.

Then Lee Iacocca came calling for Shelby's help to inject a sports-car image into the Mustang. Shelby didn't need another project. In late 1964, Shelby American's Southern California facility was bursting at the seams with projects, including phasing out the original 289 Cobra, developing the 427 Cobra, campaigning Cobra Daytona Coupes in the World Manufacturers' Championship, and taking on the task of building Ford GT40s into Ferrari beaters. But the Mustang was important to Ford, and staying in

above: Like many original Shelby Mustang owners, Hunt Palmer-Ball added "Cobra" aftermarket equipment, in his case the Shelby dual-quad intake with a pair of 460 cfm Holleys. *Juan Lopez-Bonilla*

top: Carroll Shelby with his Cobras, street and race. *Shelby American*

opposite: With the 1965 GT350 added to the Cobra and race car workload, Shelby American relocated to a pair of huge hangars near Los Angeles International Airport, where an assembly line was constructed to expedite production. *Shelby American*

Ford's good graces was important to Shelby. The Shelby GT350, and later the big-block GT500, would secure Shelby's legend in automotive history.

In 1967, Hunt Palmer-Ball chose a competition path for his new GT350. But instead of road racing, he prepared his 1967 Shelby for drag racing, eventually running 7.40-second ETs in the eighth mile and "pulling the front wheels 4 inches off the ground!" Palmer-Ball still owns his Shelby today, restored to the way he bought it new in 1967.

1963–65 Cobra

Carroll Shelby's Cobra was not your typical muscle car of the early 1960s. Nor was it purchased by the typical muscle car buyer. With a nearly $6,000 price tag, the two-seater Cobra sports car cost more than a Corvette but came with no amenities; air conditioning and power steering were not offered, and the interior door handles were simple pull straps. Raw, powerful, and exclusive, the Cobra appealed to upscale professionals, not the dirty-fingernail, white–T-shirt hot-rodders who were street racing 406 and 427 Galaxies.

Shelby's first Cobras were produced in mid- to late 1962, with the initial seventy-five or so powered by Ford's 260-cubic-inch two-barrel V-8, an adequate powerplant for a small car with a curb weight of barely more than 2,000 pounds. *Mechanix Illustrated's* Tom McCahill described it as "a hairy-chested, swashbuckling little rat

By 1965, the Cobra had evolved with 289 engine, rack-and-pinion steering, and more reliable electronics, but it was still a basic V-8-powered sports car with few amenities other than speed, handling, and exclusivity. *Jerry Heasley*

above: The first Cobras were powered by the 260 two-barrel engine. This is the CSX2000, the first Cobra, which remained in Carroll Shelby's possession until after his death in 2012. *Jerry Heasley*

right: Carroll Shelby maintained his first Cobra with its original 260 V-8 engine. *Jerry Heasley*

The rich and famous flocked to Shelby's Cobra. Here, Carroll Shelby shows off a 289 model to actor Steve McQueen.
Shelby American

that will snap Gramp's head right off his shoulders." With the arrival of the Fairlane's 289 High Performance in early 1963, Shelby American quickly switched to the 271-horsepower small-block with solid lifters and four-barrel carb, giving birth to the legendary 289 Cobra.

Sports Car Graphic tested a 1964 street Cobra with driving duties handled by Jerry Titus, who would soon become a team driver for Shelby American. Staffer John Christy wrote,

> "[The Cobra] will double the usual highway speed limit, and it will do it so fast that it is almost unbelievable. Torque is such that one merely picks an intermediate gear, instantly reaches the desired speed, and then drops it into high gear, be that 20 or 120 mph. It is also totally controllable, so much so as to be unbelievable."

Christy also noted the Cobra's shortcomings, describing personal comfort as "adequate" and weather protection as "minimal" but went on to justify the Cobra as "a sports car, one of the best in the world, and it doesn't pretend to be anything else."

The Cobra also excelled as a race car, winning numerous SCCA championships in 1963–64, both as factory Shelby entries and customer cars.

The small-block Cobra evolved over its three-year production span, upgrading over time to rack-and-pinion steering, a larger radiator, more reliable Ford electronics, and side vents for improved passenger ventilation. By the time the 289 Cobra was replaced by the 427 Cobra in 1965, Shelby American had produced 655 small-block Cobras, most sold by Shelby-authorized Ford dealerships and all adding to the Shelby legend.

1966–67 Cobra 427

A 427 in a two-ton Galaxie was one thing; dropping the race-inspired big-block into an under–3,000-pound two-seater was insanity. But that's how Carroll Shelby rolled.

When reports filtered back to Shelby American that Ferrari was developing a more powerful race car for 1965, Shelby American driver and engineer Ken Miles took matters into his own hands by shoehorning a NASCAR 427 into a Cobra. After testing and refinement, it was clear that the big-block engine was the wave of racing's future. To legally race in the SCCA, Shelby needed to produce at least one hundred street cars. When the powerful 427 proved more than the Cobra's original English chassis could handle, Ford stepped in to design a heftier frame and a coil-spring, all-independent suspension, which steadied the Cobra's skittish handling while also hooking the big rear tires to the pavement under acceleration. The Cobra's body also grew muscles, with bulging fenders to fit wider tires and a larger grille opening to help cool the massive powerplant.

The 427 Cobra was phased in as the 289 Cobra was phased out during the latter part of 1964. Dual-quad 427s were installed in approximately the first 100 cars, but then Shelby American ran into a problem with the supply of engines from Ford. At that point, the single four-barrel 428 Police Interceptor was chosen as an adequate replacement, although purchasers weren't made aware of the 1 cubic-inch disparity. However, 427s became available again toward the end of 427 Cobra production, so the final 50 or so cars got a single four-barrel 427. Total street-car production was only 260, plus another 83 as Competition, Semi-Competition, and prototype models.

The 427 Cobra was a thing of beauty, all muscle in a sports car body.
Jerry Heasley

MOTOR TREND
TEST CAR

above: Originally conceived to power Galaxies, the big-block FE was overkill in the lightweight Cobra. Although most 427 Cobras were actually powered by a 427, this one has the mid-model 428 Police Interceptor with single four-barrel. *Jerry Heasley*

opposite: With its new coil-spring suspension, the 427 Cobra could dig in off the line for impressive quarter-mile performance. *Archives / TEN: The Enthusiast Network Magazine, LLC*

Adding another 100 or more horsepower to the already potent Cobra was crazy to the max, but no one ever said Carroll Shelby was level-headed when it came to performance. The car also came at a price: $7,000 was the starting point for the overpowered two-seater with plastic side windows and a barely usable top.

When *Car & Driver* tested a 1966 427 Cobra—listing the dual-quad big-block with 485 horsepower—its pages gushed over the no-compromise muscle machine: "The Cobra has retained its identity as a raw-boned, wind-in-the-face sports car," said the writer before describing the acceleration as "unbelievable," the brakes as "magnificent," and the overall driving experience as "amazingly tractable." On the drag strip, *C&D's* Cobra registered a 12.20-second quarter mile at 118 miles per hour, by far the quickest and fastest quarter-mile performance by any American street car of the era.

C&D also put the 427 Cobra to the 0–100–0 test, resulting in 14.5 seconds to accelerate to 100 miles per hour, then panic-brake to a complete stop. "Until something better comes along," they said, "that may have to stand as some sort of high-water mark in performance for cars that are readily available to the general public."

Sports Car Graphic coaxed a 13.2-second quarter mile out of its 427 Cobra. "If one is in either of the two lower gears, one is advised to make sure the rear end has caught a little bite before one goes beyond that first throttle detent," the review warned. "This Hawg will lay great strips of rubber if treated disrespectfully at any speed in either first or second, and will even do it in third gear if it is the least bit out of shape."

1965–66 Shelby GT350

With the Total Performance marketing campaign in full swing by 1964, Ford's Lee Iacocca saw the opportunity to bolster the Mustang's image as a road racer and asked for Carroll Shelby's help. After conferring with the SCCA, Shelby learned that he could convert the 289 Hi-Po fastback into the sanctioning body's definition of a "sports car" by removing the rear seat, effectively making it a two-seater, and adding mostly bolt-on performance and suspension parts to improve acceleration and handling. At least one hundred street versions had to be "series-produced with normal road touring equipment" to legalize the car for SCCA B/Production racing.

Several names were considered and hotly debated for the Shelby Mustang, including "Mustang Cobra" and "Cobra Mustang Gran Sport." During a meeting with Ford at Shelby American, Carroll Shelby ended the frustration by asking an employee to estimate the distance between the production and race shops. When he answered, "About 350 feet," Shelby replied, "That's what we'll call it—GT350." And the matter was settled, with Shelby adding, "If it's a good car, the name won't matter. And if it's a bad car, the name won't save it."

Carroll's Super Snake

Reportedly, Carroll Shelby was driving a 427 Cobra when he lost a race against his attorney's Ferrari, leading Shelby to commission the build of an "unbeatable" 427 Cobra. Starting with a Competition model, the Shelby American crew installed a 427 with aluminum Medium Riser heads and twin Holley four-barrels atop a cross-ram intake. Then they bolted on a pair of Paxton superchargers, each delivering six psi of boost. Shelby claimed 800 horsepower. "That might be stretching it a bit," said *Road & Track* after a drive. "Perhaps one should just fall back on the old Rolls-Royce answer and say 'adequate.'"

Shelby's dual Paxton 427 Cobra, which became known as the *Super Snake*, took muscle car to the extreme. A huge hood scoop provided both free breathing and breathing room for the massive 427. Fearing that no clutch disc of the era could reliably handle the power, the 427 was backed by a C6 automatic.

Road & Track took Carroll's personal Cobra to the drag strip where the 2,500-pound sports car laid down an 11.86-second quarter mile at 115.5 miles per hour. "Mind you, that was achieved without wheel spin," they reported. "The blue Cobra doesn't get off the line like a cannon shot, but when those blowers start whining . . ."

Shelby's twin Paxton Cobra was completed as the last of the 427 Cobras were assembled at Shelby American. A duplicate was built for comedian Bill Cosby, whose short-lived ownership resulted in a standup routine highlighted on the album *200 M.P.H.*

above: With a pair of Paxton superchargers feeding the twin Holley carburetors, Carroll Shelby claimed 800 horsepower from the Super Snake's massive 427. Comedian Bill Cosby described it as having "dual everything." *Jerry Heasley*

top: *Road & Track* described Carroll Shelby's twin Paxton *Super Snake* monster as "the Cobra to end all Cobras." *Jerry Heasley*

The 1965 Shelby GT350 was essentially a road-race Mustang for the street. *Eric English*

With Ken Miles handling the development work and Chuck Cantwell hired as project manager, Shelby American created its recipe for the GT350. Starting with semi-complete 289 High Performance Mustang fastbacks, all Wimbledon White, Shelby met the SCCA's specifications by replacing the rear seat with a fiberglass panel that also mounted the spare tire. Using a combination of aftermarket and available Ford parts, some installed at Ford's San Jose, California, assembly plant and others added by Shelby, the Mustang's handling capabilities were enhanced with a one-piece cowl-to–shock tower brace (known as an "export brace" for its use on exported Mustangs), a "Monte Carlo" bar across the front of the engine compartment, A-arms lowered by 1-inch, over-ride traction bars, and 15-inch steel wheels, painted silver, with Goodyear Blue Dot tires.

Under the hood, Shelby started with the 271-horsepower 289 High Performance, then boosted the output to 306 with a high-rise aluminum intake, 715 cfm Holley four-barrel, and Tri-Y headers that dumped into glasspack mufflers with pipes exiting in front of the rear tires. Finned aluminum "Cobra" valve covers and 6½-quart oil pan were also added. For the drivetrain, Shelby specified the Borg-Warner aluminum-case T-10 four-speed and 9-inch rear end with 3.89:1 gears, Detroit Locker differential, and larger station wagon rear drum brakes.

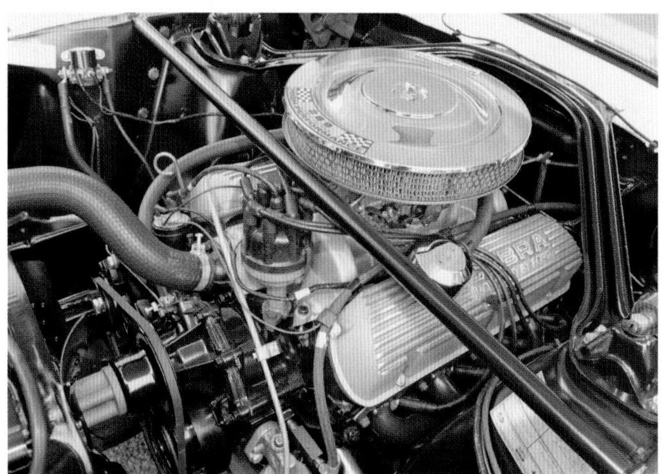

Externally, the GT350s received a fiberglass hood with open scoop and hood pins, blue side stripes, and a small running-horse emblem in the grille, offset to the driver's side and replacing the Mustang's large chrome horse and bars. Inside, the Shelby Mustangs came with competition seatbelts, three-spoke aluminum steering wheel with wood rim, and a plastic pod mounted on the center of the instrument panel, with 8,000-rpm tach and oil pressure gauge. Shelby also offered optional Cragar five-spoke aluminum wheels and, mainly as an incentive to dealers, the over-the-top Le Mans stripes in Guardsman Blue that would become a Shelby staple over the ensuing years.

The GT350 was a new type of American performance car. Unlike the 427 Galaxie or Hi-Po Fairlane, the hopped-up Shelby Mustang was built as much for handling as it was straight-line acceleration. While other muscle cars of the era were delivered straight from the factory, the GT350s detoured through Shelby American for their makeover before being sold through authorized Ford dealers. They were loud, rough-riding, and an absolute thrill for enthusiast drivers.

opposite top: In base form, the 1965 Shelby GT350 was delivered with steel 15-inch wheels and side stripes only. *Jim Smart*

opposite bottom left: Shelby American coaxed an additional 35 horsepower from Ford's 289 High Performance with Holley carb, aluminum intake, and headers. Finned aluminum "Cobra" valve covers were also part of the package. *Jerry Heasley*

opposite bottom right: By covering the rear seat area with a fiberglass panel, Shelby created a two-seater "sports car" in the eyes of the SCCA. It was also used to mount the spare tire. Competition seatbelts were part of the 1965 GT350 package. *Jerry Heasley*

right: "The GT350 is one car that will never put you to sleep at the wheel," said *Motor Trend*, after flying high in its test car. *Archives / TEN: The Enthusiast Network Magazine, LLC*

Preceded by Shelby's reputation with Cobras, the car magazines couldn't wait to test the Shelby Mustang. "The GT350 is pretty much a brute of a car," summarized *Road & Track*. *Car & Driver* described it as a "brand-new clapped-out race car." *Mechanix Illustrated's* Tom McCahill put his own twist on Shelby's Mustang: "[The GT350] is one of the greatest American Gran Touring cars ever offered in this price range. It has looks, high performance, and the road-and-dart ability of a gazelle."

At a list price of $4,547—more than $2,000 over the cost of a base Mustang—Shelby American sold 506 street GT350s for 1965, well over the 100 required by the SCCA. Total production was 562, including 36 competition models that helped attain Iacocca's goal of winning the SCCA's B-Production championship.

Although the 1965 GT350 sold better than expected, Ford saw the opportunity for even more sales in 1966 by taming the Shelby Mustang to appeal to a wider base of potential buyers. The loud side-exit exhaust and locking rear end were eliminated, the back seat stayed in, and four additional color choices were added instead of only 1965's Wimbledon White. To save production time and cost, the over-ride traction bars were replaced by under-ride versions, and the labor-intensive lowered upper control arm operation was discontinued.

To differentiate the latest Shelby aside from the 1966 Mustang's new horizontal bar grille, Plexiglas windows replaced the grille work in the C pillar, and functional brake cooling scoops were added in front of the rear wheelwells. Over time, the 15-inch wheels were replaced by 14-inch, either Magnum 500s or optional ten-spoke aluminum. Because the 1966 Mustang's new five-pod instrument panel included an oil pressure gauge, the special dash pod from 1965 was eliminated, replaced by an 8,000 rpm tach bolted to the dash pad. The Cobraized 289 High Performance engine stayed the same with its 306-horsepower rating, but the 1966 availability of the Cruise-O-Matic transmission for the Hi-Po made it possible to shift automatically.

inset: By 1966, Carroll Shelby was proud to pose with his GT350 and a 427 Cobra for Shelby American advertising. *Donald Farr Collection*

top: To broaden the 1966 GT350's appeal, it was offered in five colors and differentiated from the standard Mustang fastback with rear quarter windows and side brake-cooling scoops. This one has the optional ten-spoke aluminum wheels. *Eric English*

Supercharged GT350

For 1966, Shelby considered offering a Paxton supercharged GT350 as a special GT350S model but later decided to simply offer the blower as a $670 option. In advertising, Shelby claimed a 46 percent power increase, taking the Cobra 289 from 306 horsepower to somewhere between 390 and 400. Only eleven were built at Shelby American, although others were dealer-installed. *Car Life* tested a supercharged GT350, calling it "superior to any other Mustang we have tested" after running a 14.0-second quarter mile.

The Paxton option continued for the 1967 GT350 with thirty-five produced.

above: Shelby American considered offering a supercharged Shelby Mustang as a separate "GT350S" model. Only one was built as a prototype before deciding to make the Paxton an option for the regular GT350. *Jerry Heasley*

below: Paxton's centrifugal supercharger used a special carburetor enclosure for pumping the boost into the engine. *Jerry Heasley*

CALIFORNIA
2Y MFG 013

The taming of the GT350 worked as planned. Shelby production reached 2,378 for 1966, including 1,000 cars for a Hertz rental car program and four specially built convertibles.

1966 Shelby GT350H

Shelby American was on track to more than double the sales of the GT350 when it received a "gift" from Hertz Rental Cars. At first Hertz implied an order of 500-700 cars, all in the Hertz black and gold colors, for its Hertz Sports Car Club, where prequalified renters could pilot a GT350H for thirteen dollars a day and thirteen cents a mile. Later, the order was upped to one thousand with white, blue, and red cars mixed in but still with the gold stripes. Famously, some Hertz GT350s were rented specifically for weekend track duty, then returned on Monday. There were even stories of renters yanking the Cobra 289 engine and replacing it with a standard 289 before returning the car to Hertz.

The rental-car availability with an automatic piqued the interest of the car magazines. After testing a GT350H, *Car & Driver* remarked, "The changes for 1966 have made the GT350 more civilized, and we still think it's a great sports car in the classic tradition." At *Sports Car Graphic*, technical editor and by-then Shelby team driver Jerry Titus complimented the Cruise-O-Matic: "The three-speed automatic is Ford's pride and joy. It's a good gearbox with positive upshifts and good ratios that complement the accelerative ability of the 289 High Performance engine."

opposite: *Sports Car Graphic*'s Jerry Titus was impressed with the GT350H's automatic transmission: "The shift is instantaneous, so you need only anticipate the redline on the tach mounted atop the dash by 150 rpms and keep your throttle foot flat to the floor." *Archives / TEN: The Enthusiast Network Magazine, LLC*

below: Shelby produced 1,000 GT350H Mustangs for the Hertz Sports Car Club rental car program. Most were painted in the Hertz corporate colors of black with gold. *Jerry Heasley*

1967 Shelby GT500

Thanks to Ford's redesign for 1967, the Shelby Mustang changed in two major ways—more extensive use of fiberglass provided a totally different look than the standard Mustang fastback, and the addition of the 390 allowed Shelby to add a new GT500 model with a big-block, only bigger with a 428.

With its premium sticker price over the Mustang, Ford wanted a more individual look for the 1967 Shelby, a process that started in 1966 with Plexiglas quarter windows and side scoops. However, the new-for-1967 styling left no doubt that the Shelby was a different kind of Mustang. In fact, it was hardly recognizable as a Mustang with its use of fiberglass for the front nose, hood with scoop, rear end with ducktail spoiler, and two scoops on each side. At first, high-beam headlights were positioned in the center of the grille. When it was learned that they were illegal in some states, alternate outboard versions were devised, which also alleviated overheating issues for cars with air conditioning and/or delivery to warm-weather climates. At the rear, the 1967 Shelby was distinguished by horizontal Cougar taillights.

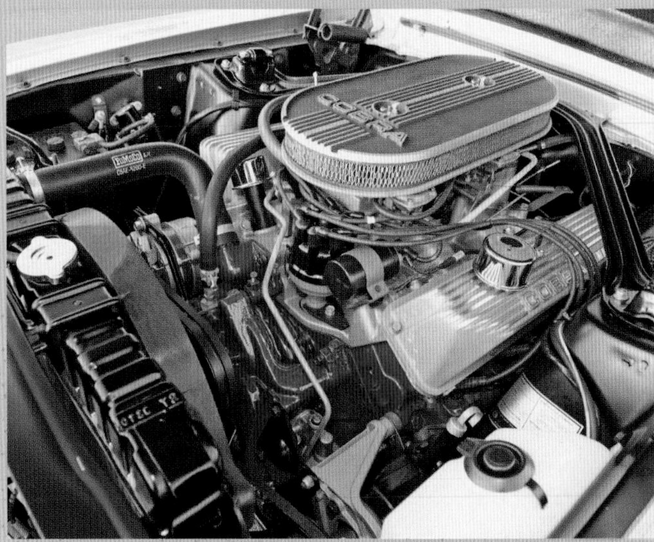

The Shelby interior started as Mustang Deluxe with brushed aluminum trim, plus the factory in-dash 8,000-rpm tach and 140-miles-per-hour speedometer. To that, Shelby added a wood-rimmed steering wheel and a pair of gauges—oil pressure and ammeter—mounted under the dash in a metal bezel, which was actually an upside-down 1966 Mustang Rally-Pac housing. Notably, the 1967 Shelby interior was also equipped with the first-time use of a roll bar that also incorporated inertia-reel shoulder harnesses.

From Ford, the fastbacks destined for Shelby were equipped with the Mustang's Competition Suspension, which was further upgraded by Shelby with progressive rate springs and larger front sway bar. Wheels returned to 15 inches, using steel rims with Thunderbird hubcaps as standard equipment and either

top: The 1967 Shelby's wide rear taillights were borrowed from the 1967 Cougar. *Donald Farr*

above: The GT500's engine wasn't just any 428 Police Interceptor. Shelby's big-block came with an aluminum intake and a pair of 600 cfm Holley four-barrel carbs. *Donald Farr*

Kelsey-Hayes five-spoke or aluminum ten-spoke wheels as options.

While the GT350 continued with the high-revving, 306-horsepower Cobra 289 High Performance, the GT500 stepped up from the Mustang's 390 with a 428 Police Interceptor topped by an aluminum Medium Riser intake and a pair of Holley four-barrels for 355 horsepower. With finned aluminum oval air cleaner and tall "Cobra Le Mans" valve covers, the 428 was an impressive sight under the fiberglass hood. Both four-speed and automatic were available, as was air conditioning, for the first time in a Shelby.

The 1967 Shelby was a transition model between the rough-and-ready earlier cars and the more luxurious Shelbys to come. Shelby promoted them as the "Road Cars," suggesting more Gran Touring than B-Production racing, a fact not lost on *Car & Drive* when it described the 1967 GT500 as "a grown-up sports car for smooth touring." The review continued, "No more wham-bam, thank-you-ma'am, just a purring, well-controlled tiger."

The new looks and available big-block model resulted in sales of more than 3,000 1967 Shelby Mustangs, with the GT500 outselling the GT350, 2,044 to 1,135.

above: The GT500's engine wasn't just any 428 Police Interceptor. Shelby's big-block came with an aluminum intake and a pair of 600 cfm Holley four-barrel carbs. *Donald Farr*

below: *Motor Trend* had no trouble smoking the rear tires during its drag test of the 1967 Shelby GT500. *Archives / TEN: The Enthusiast Network Magazine, LLC*

Buy it . . . or watch it go by

King of the Road!

Carroll Shelby has pulled the trick of the year. He's combined Ford's new *drag champion* 428 Cobra Jet engine with his *complete* road car, the Cobra GT 500. Result? Cobra GT 500-KR . . . King of the Road.

Drag champion engine? The 428 Cobra Jet grabbed Super Stock Eliminator honors at the Pomona Winternationals. It delivers 335 hp at 5400 rpm, churns up 440 lbs/ft of torque at a usable 3400. Look for 0 to 60 times that will snap your eyeballs! "Hot Rod" Magazine calls it ". . . the fastest-running Pure Stock in the history of man."

The complete Shelby Cobra GT is ready-made for the "all-there" Cobra Jet. Power is controlled by adjustable shocks, heavy-duty suspension, four-speed transmission (with automatic a low cost option), beefy driveline and torque-sensitive locking rear. All standard—along with 16-to-1 ratio power steering, high performance tires, power disc front brakes. These essentials—plus safety features, luxury interiors and limited-edition styling—are engineered-in, not just offered as options.

The game is Follow-the-Leader. The name of the game is Cobra GT 500-KR. Or play a slightly tamer game with the Cobra GT 350. But make your play at your Shelby dealer . . . today.

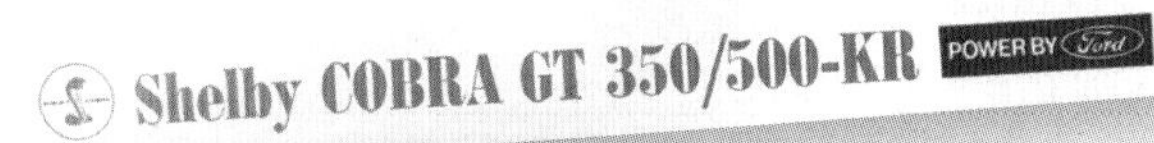

above: Shelby's "King of the Road" advertising for the 1968 GT500KR described the new CJ-powered Mustang as Carroll Shelby's "trick of the year." *Donald Farr Collection*

right: With the 428 Cobra Jet replacing the 428 Police Interceptor at mid-1968, the CJ-powered Shelbys were renamed GT500KR. Revised 1968 fiberglass included a hood with forward-placed scoops. *Jerry Heasley*

1968 Shelby GT500KR

For 1968, the Shelby GT350/GT500 received a mild exterior makeover with new fiberglass nose, hood with forward-placed scoops, Thunderbird sequential taillights, and wheels covered by faux mag-style hubcaps. Once again, the interior was the Deluxe Mustang, only with woodgrain plus a specific Shelby console. The big news was the availability of a convertible model with a styled roll bar. With Shelbys becoming more Ford and less Shelby, production moved to the A.O. Smith Company in Ionia, Michigan.

Continuing with more luxury and less performance, the 1968 GT350 was powered by a 250-horsepower four-barrel 302, and the GT500's motivation was reduced to a single four-barrel 428 Police Interceptor rated at 360 horsepower. However, when Ford introduced the 428 Cobra Jet in

177352
WISCONSIN

left: Although rated lower than the previous 428 PI, the KR's midyear 428 Cobra Jet was a performance engine capable of quicker acceleration. Ram-Air was included. *Jerry Heasley*

below: With extensive use of fiberglass, the 1967 GT500 looked more like a Shelby and less like a Mustang. This one has the center-mount high-beam headlights, which were illegal in some states, resulting in a second grille design with outboard lights. *Donald Farr*

The GT500KR was available in the new Shelby convertible for 1968, complete with roll bar. Many owners replaced the stock hubcaps with Shelby's ten-spoke aluminum wheels. *Eric English*

April 1968, the new engine became the standard GT500 powerplant with functional Ram-Air. Although the CJ's 335-horsepower rating was less than before, it was an all-out performance engine built for brute acceleration. To differentiate the CJ cars, "KR" was added to the GT500 lettering for "king of the road," a name intended for a high-performance Corvette until Carroll Shelby swooped in to snatch the copyright from under Chevrolet's nose.

The 428 Cobra Jet put the 1968 Shelby GT500KR into supercar territory capable of low to mid 14-second quarter miles. *Car Life* said:

> "The primary barrels open and nothing happens, then whump—the power comes on and the rear end slides out. Correct with the steering, and whump—more power, more slide, more correction. Then the secondaries, controlled by vacuum and not by the driver, open and whump—still more power, slide, and correction. Floor the gas pedal. It's about all the enjoyment the driver can stand."

Total 1968 Shelby production reached 4,451, including 1,570 as GT500KRs with the 428 Cobra Jet.

1969–70 Shelby GT500

By the time the 1969 Shelby arrived in Ford dealerships, the word was out: Shelby was done. Ford had the Mach 1, Boss 302, Boss 429, and even a Fairlane with the Cobra name, so there was no need for a Shelby in the mix. "This year the enthusiast bought a Mach 1 or a Boss," noted *Car Life*. "Why spend the extra $800 for a Shelby?" The 1969 GT350s and GT500s would be the last of the Shelby Mustangs, although a few leftovers would be reassigned as 1970 models.

Like the Mustang it was based on, the 1969 Shelby was larger and heavier. Yet, even more than 1967–68, the new Shelby looked less like a Mustang. Offered as a

GT500 Super Snake

As a longtime Goodyear West Coast distributor, Carroll Shelby was asked to participate in a high-speed demonstration for a new tire at Goodyear's Texas test track. It seemed like the perfect opportunity to install a 427 into a GT500, with Shelby American sales manager Don McCain suggesting a build of fifty as a special *Super Snake* model. At Shelby American, a Medium Riser 427 was prepared with aluminum cylinder heads, solid-lifter cam, headers, and 780 cfm Holley four-barrel, which was installed into a white 1967 GT500 along with a Toploader four-speed and 4.11:1 gears. Blue Le Mans stripes, in a triple narrow-wide-narrow pattern instead of the usual dual striping, set the *Super Snake* apart from standard GT500s.

During the high-speed test, the *Super Snake* reached 170 miles per hour and set a record with a 142-mile-per-hour average for the 500-mile session. However, the *Super Snake* never materialized as a special model for customer consumption when it was realized that the cost of the 427 would push the sticker price to more than $7,500.

left: The 1967 *Super Snake* at speed on Goodyear's test track in Texas. It reportedly reached a top speed of 170 miles per hour. *Shelby American*

below: Shelby built a special 427 for the *Super Snake*'s Texas tire test. *Al Rogers*

convertible or SportsRoof fastback, the entire front clip was fiberglass, resulting in a full-width grille. The hood incorporated five recessed and functional ducts; the front three were NACA ducts with the center opening funneling cold air to the engine and the outboard pair supplying cooler air to the engine compartment. The rear-facing openings extracted warm under-hood air. Rear brake-cooling scoops were added to the rear quarters. Mid-mounted side-stripe decals represented one of the first uses of light-reflective material on a production car.

At the rear, a fiberglass trunk lid and quarter panel caps created an even bolder ducktail spoiler than the production Mustang fastback, with wide, sequential-operation 1965 Thunderbird taillights tucked into the rear panel. Totally new and different for the 1969 Shelby was the cast aluminum exhaust outlet mounted under the center of the rear bumper.

The Mustang Deluxe interior supplied woodgrain trim, with an 8,000-rpm tach and 140-mile-per-hour speedometer included. For exclusivity and function, the Shelby console incorporated a pair of gauges (oil pressure and amp) and toggle switches to operate the Lucas driving lights and interior illumination. The roll bar returned for both fastbacks and convertibles.

Thankfully, the GT350 got more power from the 351 Windsor, Ford's new, larger-displacement small-block. Although equipped with a Shelby aluminum intake manifold, 470 cfm Autolite four-barrel, and functional Ram-Air, it kept Ford's factory rating of 290 horsepower. Buyers looking for maximum performance chose the GT500 with

The 1969 Shelbys were a radical departure from the 1968s—and from the Mustangs they were based on. Designed at Ford, the fiberglass front end previewed the look of the 1971 Mustang. *Jerry Heasley*

SHELBY

above: One of the most notable features for the 1969–70 Shelby was the center-mounted exhaust port. *Jerry Heasley*

left: Faced with 788 leftover 1969 Shelbys, Ford simply added hood stripes and a chin spoiler to differentiate them as 1970 models. *Jerry Heasley*

R-code 428 Cobra Jet, which was identical to the Mustang's CJ with 335 horsepower. Like the Mustang, the Shelby's Cobra Jet was available with the Drag Pack option, which supplied 3.91 or 4.30 gearing and upgraded the engine to a Super Cobra Jet with oil cooler and heavy-duty rotating assembly. The suspension was Mustang's Competition version with staggered rear shocks supplied for four-speed cars, along with new Shelby exclusive 15x7 five-spoke aluminum wheels and available F60 Goodyear Polyglas tires.

Although beautiful and fast, the 1969 Shelby couldn't escape comparisons to the original Shelby Mustang from 1965–66. "A garter snake in Cobra skin," said *Car & Driver*. It described the 1969 model as "sort of a baby Thunderbird—a Turnpike Cruiser with slots."

Car Life gave the 1969 GT500 its full road-test treatment, complimenting the suspension's solid feel and the well-located instruments, along with describing the exterior as "beautiful as Raquel Welch." After posting an impressive fourteen-seconds flat quarter mile, the editors backed away from the comparisons to 1965: "We were pleased to see the Shelby take on an image of its own. It's a beautiful thing. Those bulging lines you see from the driver's seat give you a sense of power and a stomp on the throttle pedal will tell you it's not just wishful thinking."

With in-house competition from the Mach and Boss Mustangs, Shelby sales dropped to 2,362 for 1969, more than half as GT500s. In fact, Ford ended up with 788 leftovers, which were updated into 1970 models with hood stripes, front chin spoiler, and revised VIN.

BY THE NUMBERS:

1964 289 COBRA

ENGINE: Cobra 289 High Performance

CARBURETION: Autolite four-barrel

HORSEPOWER: 271 at 5,800 rpm

TORQUE: 312 at 3,400 rpm

TRANSMISSION: Four-speed

REAR AXLE RATIO: 3.78:1

WEIGHT: 2,310 lbs (test)

HORSEPOWER TO WEIGHT: 8.5

QUARTER MILE: 14.9 at 93 mph (*Sports Car Graphic*, November 1963)

1966 427 COBRA

ENGINE: 427

CARBURETION: Two Holley four-barrels

HORSEPOWER: 485 at 6,000 rpm

TORQUE: 480 at 3,700 rpm

TRANSMISSION: Four-speed

REAR AXLE RATIO: 3.54:1

WEIGHT: 2,890 lbs (test)

HORSEPOWER TO WEIGHT: 5.95

QUARTER MILE: 12.2 at 118 mph (*Car & Driver*, November 1965)

1965 SHELBY GT350

ENGINE: Cobra 289 High Performance

CARBURETION: 715 cfm Holley four-barrel

HORSEPOWER: 306 at 6,000 rpm

TORQUE: 329 at 4,200 rpm

TRANSMISSION: Borg-Warner four-speed

REAR AXLE RATIO: 3.89:1 Detroit Locker

WEIGHT: 3,140 lbs (test)

HORSEPOWER TO WEIGHT: 10.26

QUARTER MILE: 14.7 at 90 mph (*Road & Track*, June 1965)

1966 SHELBY GT350H

ENGINE: Cobra 289 High Performance

CARBURETION: 715 cfm Holley four-barrel

HORSEPOWER: 306 at 6,000 rpm

TORQUE: 329 at 4,200 rpm

TRANSMISSION: Three-speed automatic

REAR AXLE RATIO: 3.89:1

WEIGHT: 3,158 lbs (test)

HORSEPOWER TO WEIGHT: 10.32

QUARTER MILE: 15.2 at 93 mph (*Car & Driver*, May 1966)

1967 SHELBY GT500

ENGINE: 428 Police Interceptor

CARBURETION: Dual Holley 600 cfm four-barrels

HORSEPOWER: 355 at 5,400 rpm

TORQUE: 420 at 3,200 rpm

TRANSMISSION: Four-speed

REAR AXLE RATIO: 3.25:1

WEIGHT: 3,576 lbs (test)

HORSEPOWER TO WEIGHT: 10.07

QUARTER MILE: 14.3 at 92 mph (*Sports Car Graphic*, March 1967)

1968 SHELBY GT500KR

ENGINE: 428 Cobra Jet

CARBURETION: Holley four-barrel

HORSEPOWER: 335 at 5,200 rpm

TORQUE: 440 at 3,400 rpm

TRANSMISSION: Four-speed

REAR AXLE RATIO: 3.50:1 Traction-Lok

WEIGHT: 3,780 lbs (test)

HORSEPOWER TO WEIGHT: 11.28

QUARTER MILE: 14.57 at 99.55 mph (*Car Life*, October 1968)

1969 SHELBY GT500

ENGINE: 428 Cobra Jet

CARBURETION: Single Holley four-barrel

HORSEPOWER: 335 at 5,200 rpm

TORQUE: 440 at 3,400 rpm

TRANSMISSION: Automatic

REAR AXLE RATIO: 3.50:1

WEIGHT: 4,230 lbs

HORSEPOWER TO WEIGHT: 12.62

QUARTER MILE: 14.0 at 102 mph (*Sports Car Graphic*, February 1969)

INDEX

About the Author

Donald Farr has been writing about Mustangs and Fords for over 40 years. Over that time, he has edited many of the Ford-related magazines, including *Mustang Monthly*, *Super Ford*, *Mustang & Fords*, and *Mustang Times*. He was also the founding editor of *Musclecar Review*. Farr is also the author of *Mustang Boss 302: From Racing Legend to Modern Muscle Car*, *Art of the Mustang*, and *Mustang Fifty Years*, which has been revised into *Ford Mustang: America's Original Pony Car*. He is a Lee Iacocca Award recipient and 2012 inductee into the Mustang Hall of Fame.